to Mike
Best Wishes

Ti 10/09

Just A Little About Most Everything

Essays on Technology and Life

Tim Wooldridge

Copyright © 2009 by Tim Wooldridge

ISBN 0-7414-5476-9

Cover graphics by Meghan List. Copy editing by Michelle Heeden King. Photos with permission from iStockPhoto, NASA, and Tim Wooldridge.

Published by:

INFINITY
PUBLISHING.COM

1094 New DeHaven Street, Suite 100
West Conshohocken, PA 19428-2713
Info@buybooksontheweb.com
www.buybooksontheweb.com
Toll-free (877) BUY BOOK
Local Phone (610) 941-9999
Fax (610) 941-9959

Printed in the United States of America

Published July 2009

Acknowledgments

The author thanks Barbara McRae, editor extraordinaire of the *Franklin Press*, and Buchmann for encouragement and suggestions.

Dedication

This book is dedicated to my family and particularly to Grandad (1910-2007), whose inspiration continues.

Table of Contents

Topics in Astronomy and Cosmology

The Lore of the Moon

Ever hear the phrase, "once in a blue moon?" Have you heard someone talk about a harvest moon or a waxing gibbous moon? Since the moon is the most impressive object in the night sky, an interesting collection of terms has developed. Mixed in with the terms is plenty of lore about our moon – some true, some not.

The common definition of a blue moon is the second full moon that occurs in one calendar month. This is not a rigorous astronomical definition because astronomers don't much care about calendar months. Because of the moon's orbit around the Earth, we have a full moon every 29.5 days. That makes it quite possible to have two full moons in one month. The phrase "once in a blue moon" usually means something rare, but blue moons are not really rare. The average occurrence is once every 32 months. No one seems to know why the term "blue" was applied to this phenomenon. The color of a full moon is really determined by atmospheric conditions. Smoke or dust in the air can alter the color from blue through white to orange.

The phases of the moon have some curious names. Full moon and new moon are common terms, but the whole sequence is:

New moon	The moon is not illuminated on the side we see
Waxing crescent	A sliver, illumination is from the right side
First quarter	Half-illuminated from right side
Waxing gibbous	Between half and full, still from the right
Full	Fully illuminated

Waning gibbous	Between half and full, illuminated from the left
Third quarter	Half illuminated from the left side
Waning crescent	A sliver, illumination is from the left side
New Moon	Again

The full moons have names given to them by activities of the time of year. For example, the full moon in October is often called a harvest moon.

What about the other stories of epileptic seizures, madness, and wolves howling during the full moon? Sorry. No correlation. Wolf vocalization is an important communication tool in a pack. Howling at the moon would be a waste of their breath.

Have you heard or seen that the full moon seems much larger when it is first rising than when it is overhead? If you measure it carefully, you will find that it is not. This illusion is dramatic, but it is due to the fact that you have reference points when the moon is near the horizon and none when it is overhead. Your mind perceives the moon as being larger when you have those reference points available.

How about the dark side of the moon? Actually, the moon doesn't have a dark side. It rotates on its axis like the Earth does, only more slowly. A day on the moon is about 27 Earth days. This produces illumination of both sides by the Sun. It happens that this axis rotation is matched to the moon's orbital motion around the Earth so that the same side of the moon is visible all the time. This synchronization of rotation with orbital motion is fairly common with natural satellites.

Where did the moon come from? The scientific debate continues on this subject. A leading theory proposes a collision with the Earth from a Mars-size object early in the formation of the Earth. The debris is believed to have coalesced into the moon.

When you look at the moon with a small telescope, the number of craters is startling. Why so many? The answer might be even more startling. The Earth likely has experienced just as many impacts. Ours have just been covered up by wind and water, neither of which the moon has. It makes you hope that all that crater formation is over with, but that's another subject.

How's the Weather Up There on the Sun?

It doesn't seem like the fiery, nuclear furnace that is our Sun could have weather, but it does.

Our Sun not only has a major effect on the weather here on Earth but also has its own sort of weather. At least, the Sun has events that are akin to storms. These stormy disturbances are all related to changes in the distribution of the Sun's magnetic field.

One type of disturbance produces dark spots, called sunspots, which can be seen on the surface of the Sun. (*Important*: Never look directly at the Sun. Safe Sun viewing is possible only by projecting the image of the Sun or by using filters designed specifically for viewing the Sun.) A sunspot is the solar equivalent of a hurricane. They were observed and recorded as early as the 1600's by several astronomers, including the well-known Galileo.

During times when sunspots are present, many types of radio communications on Earth are affected. Some short-wave radio communications improve dramatically and allow communications at very long distances. The improvement comes from changes to the Earth's ionosphere. This multi-layered structure high in the Earth's atmosphere can reflect short-wave radio signals. Areas around sunspots emit increased amounts of extreme ultraviolet (EUV) radiation that improve the ability of the ionosphere to reflect short-wave radio waves. Multiple bounces from antenna to ionosphere allow very long distance short-wave radio communications during periods of sunspot activity. At the same time, satellite communications and some other radio frequencies can be degraded somewhat.

In addition to sunspots, the Sun has sudden, explosive events called solar flares and coronal mass ejections (CME's) that have an even greater impact on the Earth. These bursts of energetic particles and radiation add to the existing "solar wind" that is constantly emitted from the Sun. When the radiation part of the enhanced solar wind reaches the Earth, the ionosphere can become so dense that some types of radio waves are absorbed, making those types of radio communication impossible.

Some of the electrons generated during these events are directed by the Earth's magnetic field into the upper levels of the atmosphere in the polar areas. As the energetic electrons crash into molecules of air, the well-known northern lights, Aurora Borealis, appear, rarely at latitudes as far south as Florida. Similar lights are visible at the South Pole.

Current flows at the top of the atmosphere can induce current spikes in electric power lines, and power outages sometimes result. (A topic for another day might be capturing and using the current flows generated by the solar wind if the electric grid were smarter.) The bursts can also cause temporary and permanent damage to the electronics in communications satellites. Even astronauts need to take special precautions if a burst occurs while they are in orbit.

How about weather forecasts for the Sun? Surprisingly, they do exist. Observations of sunspots over centuries show that the most intense periods occur approximately every 11 years. The peak periods last a few years and are followed by quiet periods of a few years that occur about midway in the 11-year cycle. The flares and CME's occur mostly, but not completely, during the periods of maximum sunspot activity.

Forecasting is also helped by the fact that radiation from these explosive events, including the visible light, travels at the speed of light. It takes about eight minutes for the light to cover the 93 million miles from the Sun to the Earth. This means that satellites and solar telescopes can detect the events within about eight minutes of their occurrence. The particles in the enhanced solar wind travel much slower than light. They can take several days to arrive at the Earth. This provides a little time to prepare. Satellites are sometimes shut down or otherwise put into safer modes until the storm passes – much like closing the shutters when a hurricane is coming.

The next few years should be some of the quiet ones on the Sun, but like those other forecasts – you know, the ones that say it's going to be clear, and it's already pouring down rain – you never know.

Do You Remember What the Stars Look Like?

If you live in the country far from any city, the title question is silly. You can look at the stars on any clear night. If, however, you live in a suburb of a large city, you probably can only see a handful of the brightest stars and planets. It is even possible for big-city kids to grow up without ever seeing stars. What happened to our night sky?

I'll give you a hint. A few years ago during a major power blackout in some eastern cities, the thing people noticed the most was the stars. Many people were seeing them for the first time in years, and some were seeing them for the first time ever.

Lights and more lights have happened to us. In orbit high above the night side of our planet, the first thing you would notice is the glow of the cities. In a startling nighttime satellite photo of the US, every city – even small ones – is visible as a blob of light.

So, we have trouble seeing the night sky because of light pollution. Some might argue that "pollution" isn't the correct term because light is useful. Outdoor lighting allows us to participate in a wide variety of nighttime activities: sports, such as football; shopping; concerts; and even driving. Outdoor lights are widely used as a crime deterrent and a safety element for both businesses and homes. Lights play a major role in advertisement. Decorative lighting is very popular, especially at Christmas. So, why call it "pollution?" Pollution can be the negative, unintended result of doing too much of something useful. Outdoor lighting falls in that category.

Is light pollution as difficult to solve as our other pollution problems? Do we have to turn off all outdoor lights? Actually, no. Almost all light pollution occurs because a large fraction of the light from most outdoor fixtures is either going straight up or over the horizon. Neither of these two directions is of any use to the owner of the light. The useful light is in a cone-shaped volume under the fixture. Any light outside that cone is wasted. The solution is not difficult to engineer. Outdoor lights need a reflective cap on top that redirects the otherwise wasted light into the cone of usefulness.

Skeptics can visit Kitt Peak, about 50 miles outside Tucson, Arizona. Kitt Peak has several major professional observatories with astronomers from all over the world. Their research work would be severely hampered by sky glow from Tucson. If you sign up for the nightly observers' program, you can see the result of local ordinances that require modified outdoor lights. Tucson is a fair-sized, well-lit city that is barely visible from the top of Kitt Peak. Phoenix, on the other hand, is a hundred miles away with no restrictions and is clearly visible, even over the horizon.

How can you tell how dark your sky is? Pick a clear night in the cooler months with no moon. Turn off all the lights that you have control over. Allow your eyes to adjust to the darkness for 15 minutes, and look for the Milky Way. This is a fuzzy band of stars that extends across the sky from roughly northeast to southwest. If you can see it, your skies are very dark. I may want to come visit you. If you can clearly see the sky itself as well as your feet, your skies are like mine and not very dark.

Have You Seen a GRB Lately?

In the world of astronomy, a great detective story has been unfolding. The perpetrator is a Gamma Ray Burst, or GRB. Like a serial criminal, a GRB appears suddenly, disappears quickly and leaves little evidence.

Gamma rays are electromagnetic radiation similar to x-rays, and GRB's are brief flashes of very intense gamma rays in the sky. They last from a fraction of a second to several hours. These intense flashes of radiation have been a long running mystery. Most events that astronomers study last much longer, often many times longer than a scientist's lifetime.

To add to the cloak and dagger of this mystery, GRB's were discovered secretly in the 1960's by US spy satellites that were monitoring the former Soviet Union for atmospheric nuclear weapons tests. The satellites were equipped with sensitive gamma ray detectors because nuclear weapons give off gamma rays at the moment of explosion. To the surprise of scientists monitoring the satellites, bursts of gamma rays were coming in frequently from random points in the sky and at random times, averaging about once per day.

For years these GRB's were frustrating events for astronomers because they didn't last long enough to be studied. When an astronomer aimed a telescope at a point in the sky where a GRB had been detected, there was nothing at all there to see. GRB's also can't really be studied from the Earth's surface because the gamma rays do not penetrate the atmosphere. All data come from orbiting satellites.

The first question to be answered was how far away are GRB's? Are they nearby in our Milky Way Galaxy or distant in other galaxies? That question was answered by the Compton Gamma Ray Observatory, launched in 1991. The bursts were confirmed to be outside our galaxy because they were not confined to the plane of the Milky Way. They were found to come from random directions in the sky. This discovery made GRB's even more of a mystery. If they are at great distances, they must represent an almost unimaginable energy release in such a short time. What are they?

Finally, in 1997, a specialized satellite spotted a GRB and quickly relayed the position information to a ground based telescope as well as the Hubble Space Telescope. Astronomers were able to see an extended object optically. Shortly after this, they were able to measure the optical spectrum of the light from another burst and calculate the approximate distance. The briefly visible object was billions of light years away. A light year is the distance that light travels in one year, so the GRB was at a great distance.

Although data are still coming in, the leading theories are that the longest lived GRB's may be explosions of especially massive stars at the end of their lifetimes. These explosions are called hypernovas. The shortest duration GRB's may be the collision and resulting instantaneous burst of energy of two burned out stars, called neutron stars. In typical scientific fashion, the debate will rage on until there are enough data to support a particular theory.

GRB's are not just of interest to astronomers. As you can imagine, you might not want to be too close to one of them when it flashes. Biological scientists are now beginning to wonder if one or more of the mass extinctions of life on Earth in the past were caused by a GRB in our galaxy. Some of these extinctions have been attributed to large meteor strikes or volcanic eruptions. A nearby GRB could destroy the protective ozone layer at the top of our atmosphere and allow the strong ultraviolet light from the Sun to destroy most plant and animal life.

Let's hope these interesting objects remain at a great distance to us. I, for one, have no desire to be a flash in the pan.

Listening to the Universe with LIGO

Astronomers and cosmologists have been looking at the universe for many years now. Innovations like the Hubble Space Telescope and several new ground-based, optical telescopes, coupled with radio telescopes like the Very Large Array in New Mexico, have provided an unprecedented view of and a vast amount of information about the universe. Scientists are struggling to develop new and more detailed theories about the origin and functioning of the universe based on this new data.

All ground-based and orbiting telescopes "look" at objects in the universe by collecting electromagnetic radiation into an image. In some cases, the telescopes use visible light, but in other cases they use invisible types of electromagnetic energy, such as radio waves, infrared light, ultraviolet light and x-rays. Up until now, none of our telescopes could detect a different type of emanation called a gravity wave.

Then along came LIGO (pronounced LYE-GO), which stands for Laser Interferometer Gravitational Wave Observatory). I don't know why they dropped the W.

To begin this amazing story, I should point out that gravity waves are theoretical. They haven't been detected in any direct way yet. They were predicted by Albert Einstein as a part of his General Theory of Relativity. According to Einstein, the changes in gravity from major astronomic events, like exploding stars, should produce distortions in space itself, which should propagate in all directions at the speed of light. These ripples in space will distort any matter as they propagate through it.

LIGO is an L-shaped structure, with each arm of the L over two miles long. In principle, the operation of LIGO is quite simple. The length of each arm is precisely measured with a laser and the lengths compared many thousands of times per second. Two LIGO installations were built, one on each coast. If a distortion is detected at the site near Baton Rouge, Louisiana, and at the Hanford site in Washington State at the same moment, a gravity wave has passed.

The principle of LIGO is simple, but the practice of getting the two detectors up and running has been difficult. The arms of the interferometer are measured with powerful infrared lasers. Air would interfere with these lasers, so the arms are contained inside steel and concrete tubes that have all of the air pumped out of them to create a high vacuum. You can probably imagine the difficulty in getting all of the leaks out of miles-long structures. Critical areas of the structure must be carefully isolated from local vibration. The presence of logging in the vicinity of the Baton Rouge site continually knocked them offline until they designed sophisticated, active vibration isolation. They can still detect waves striking shore on the Gulf of Mexico and the West Coast. Earthquakes anywhere in the world also regularly knock them offline. Although the extra vibrations are a problem, scientists are confident that they can recognize the signature of real gravity waves.

LIGO East and West are now up and running. It's probably not appropriate to say that LIGO is "looking" at the universe for a gravity "event." It might be more accurate to say that LIGO is "listening." A gravity wave will probably be in the general frequency range of sound, but sound waves, of course, can't propagate through space. To answer the obvious question: No, they haven't detected any yet.

The next question is: What if they don't detect any? The answer is typical of scientists: They increase the sensitivity of their equipment. Upgrades are already planned. Ultimately, if gravity waves still aren't detected, there may either be something wrong with our understanding of Einstein's General Theory of Relativity or the theory itself. In the world of physics and astronomy, either case would equal an exploding star.

Do You Want to See a Star Being Born?

No, not one of the short-lived Hollywood imitations; I'm talking about a *real* star. OK, it's not completely fair to say that you can see a real star being born. The whole sequence can take several million years. You can, however, see the work in progress in the Orion Nebula with a little help from binoculars or a small telescope.

At the same time that leaves are going out in their fall blaze of glory, the constellation of Orion, the Hunter, comes around in the night sky. In the late fall and early winter, Orion rises in the east and travels across the sky, remaining in the southern half of the celestial hemisphere. It is perhaps the most commonly known of the constellations, along with Ursa Major, the Big Dipper. And Orion contains two of the most interesting objects in the night sky.

You can recognize Orion by the three bright stars that define his belt. These three stars appear in a line close together between two very bright stars. From the belt stars, you can locate a bright reddish star, called Betelgeuse, at his left shoulder, and the bright blue-white star, called Rigel, at his right knee. After you have found the belt stars, imagine a short sword hanging from the belt. With your binoculars, find a fuzzy glowing spot at the tip of his sword. This is the Orion Nebula. This nebula, or glowing cloud of gas and dust, is known to be an active area of star birth.

Stars are born in a rather simple process on a grand scale. From a large cloud of hydrogen gas, a ripple may start the process of condensation. The gas is drawn together by its own collective gravity and becomes denser. As the density increases, so does gravitational pull and temperature. After a few million years, the density of the ball of gas can be high enough that it lights up in a self-sustaining nuclear fusion reaction. A star is born. The left over gas and dust in the vicinity of the new star can become planets.

The Orion Nebula is one of many large clouds of hydrogen gas. It is glowing because there are four new stars in the cloud that emit so much energy the cloud glows with visible light. These four new stars, called the Trapezium Stars, near the center of the nebula can be resolved with a good telescope.

The star at Orion's left shoulder, Betelgeuse (pronounced much like "beetle juice"), is another interesting object. Betelgeuse is a red super giant star around 600 times the size of our Sun. Its size and color indicate that it has mostly burned out the hydrogen in its core and is at the end of its life. Betelgeuse will be collapsing in a violent death, called a supernova, before long. When it happens, it will be a sight to see. It will light up the night sky so much that shadows will be cast. It should even be visible in the daytime. Unfortunately, that supernova is imminent only in astronomical time. It may still be thousands of years away.

The death of Betelgeuse and other old stars is an important part of the cycle of life. The heavy elements, including the carbon that we are made of, originally came from the dust of a supernova.

I may not get to see the Betelgeuse supernova, unless those vitamins I'm taking are a lot better than I think.

Are Extraterrestrial Solar Systems Out There?

For a few years now, astronomers have been collecting evidence of planets orbiting stars beyond our own solar system. To begin with, this evidence has been indirect. In most cases, the central star is so bright relative to the planets and so close to them that direct imaging has been impossible. The light from the planet gets lost in the glare of the star. An alien astronomer looking at our Sun from outside our solar system would have the same problem.

Astronomers use two clever tricks to gather evidence. In some cases, the orbital plane of a large planet causes it to pass directly in front of a distant star. The planet blocks some of the light from the star and makes a tiny change in the observed brightness. These tiny, periodic changes in brightness are consistent with an orbiting planet. In other cases, the gravitational pull of a large orbiting planet causes a tiny but measureable wobble in the star. A large number of stars have been associated with planets using these two tricks. The only information about the planet that can be extracted is its mass (in the case of the wobble measurement), its diameter (in the case of the brightness measurement), and its orbit (in both cases).

The current theoretical model for planet formation, based on our own solar system, suggests that we will never be able to image an alien planet directly, that they will always be too close to the star.

Recently, that theory has been shaken by images taken by both ground based telescopes and the Hubble Space Telescope. In 2004 and 2006, an easily visible star called Formalhaut was imaged with the Hubble. Astronomers were excited to find a ring of dust surrounding the star, consistent with a solar system in the process of forming planets. Even more exciting was the discovery that one edge of the dust ring was sharp rather than blurred, indicating the effect of a large planet. With further study, a bright dot was located in the image, which moved during the two years the images were taken. This appears to be the direct image of a large planet orbiting Formalhaut at a distance much greater than that predicted by planet formation theory.

In another fascinating but puzzling case, a star with the unremarkable name of HR 8799, which sits 130 light years away in the constellation Pegasus, was studied by astronomers with ground-based telescopes. What they found were three massive planets that appear to be in orbit. The most distant planet in the group is roughly seven times the mass of Jupiter and sits 68 astronomical units away from its host star (one AU is the distance between the Earth and sun). The other two closer objects are about 10 Jupiters in size. These three objects are almost large enough to be stars themselves, but they orbit HR 8799 in a single plane, like planets inside a large dust ring.

These discoveries are probably the first in a series of such discoveries. Why now? The answer lies in much better optical capability. The Hubble orbiting outside the Earth's atmosphere has excellent imaging capability. Even though it is currently limping until its next servicing mission, Hubble is one of the best imaging resources available. Ground-based telescopes, like the Keck's on the island of Hawaii, are also a leap forward. They use a technology called adaptive optics to compensate for turbulence in the air. Astronomers are simply seeing better than ever.

Progress will continue, as the Hubble is not the only space telescope. There are more planned.

Are Black Holes Imaginary, Formidable Monsters or Innocent Neighbors?

Black holes are just intriguing enough to become science fiction mainstays and turn Steven Hawking, a British theoretician who helped describe them, into a popular figure. This is not so surprising if you consider the popular description of a black hole as a monstrous entity that devours everything that gets too close, including the light that would allow us to see it. Once beyond the event horizon of the black hole, space and time are no longer defined, and the devoured object ceases to exist with any description known to science. That's pretty scary stuff, but it's not really the whole story.

Black holes are theoretical objects that arise from Einstein's General Theory of Relativity. Essentially, they are massive objects, with masses equivalent to a large star, compressed to a tiny volume. In all stars, volume is dependent on a balance between the energy of nuclear fusion (like that in a hydrogen bomb), which tries to expand the star, and the force of gravity, which tries to collapse it. When the hydrogen runs out, gravity wins.

For a smallish star like our Sun, the collapse would probably form a white dwarf – a small, but still visible, dense object. A somewhat larger star would probably form a neutron star with a density so high that all ordinary matter is collapsed to nuclear material – neutrons. If larger yet, a star will probably form a black hole.

As a theory, a black hole would seem to be difficult to prove. Anything, including light, that approaches closer than the event horizon distance is captured. That would seem to make a black hole invisible. Luckily, as a gas cloud approaches a black hole, the atoms are accelerated by gravity. These energetic atoms tend to give off light as they circle the event horizon. The disk of high energy atoms allows astronomers to detect the presence of a black hole. The concept remains a theory, but evidence is accumulating.

The monster reputation of black holes is not really deserved. Remember, a black hole is nothing but the mass of a star at a point in space. It has no more effect on neighbors than any other star of equal mass. In general, a black hole rarely captures other stars or planets because distances among stars are vast. The interstellar gas that flows into a black hole makes an insignificant contribution to its mass. The exception is the case where a collapsed star already has a companion star in a binary system. Capture of a companion to a collapsed star can either give the burned out star enough mass to become a black hole or allow a black hole to suddenly become much larger.

For an interstellar traveler, approaching a black hole would be no larger a threat than getting too close to a star.

So far, we have described what can be called a stellar black hole. An even more interesting object is the super massive black holes that seem to be at the center of galaxies and may help give them their spiral shape. How those form and what their characteristics are is more of a mystery.

Topics in Nature

Good Science in the Fall Colors

Fall is the time of the year when leaves in North Carolina undergo a spectacular transformation that is one of nature's best displays. Do you ever wonder how a leaf can change colors so completely? Why does it happen? How does the tree know when it is time to change?

Let's start with a green leaf. Why is it green? In daylight or any source of white light, leaves absorb red and blue light. If you are not sure it makes sense that a leaf absorbs red light, try looking at a leaf with a red flashlight. The leaf will look black since none of the red light will be reflected back to your eye. If a leaf is absorbing red and blue light, the remaining component of white light that comes to your eye will be mostly green.

The chemical in leaves responsible for the green color is chlorophyll. The chlorophyll is inside tiny nodules in leaf cells called chloroplasts. A chloroplast is really a microscopic but very efficient chemical factory that combines carbon dioxide from the air, water and sunlight to make sugars and starches. A by-product formed is oxygen. The sugar and starch produced in this photosynthesis is used by the tree for its normal functions, including growth. We are thankful for the oxygen by-product so we can carry on our normal functions, including breathing.

OK, leaves are green because of chlorophyll. An important characteristic of chlorophyll is that it is rather unstable. Sunlight breaks it down, so leaves only remain green while chlorophyll is being continually replaced. The chlorophyll replacement process depends on warm temperatures, plenty of sunlight and adequate water.

Leaves often have another type of chemical compound in them called carotenoids. Carotenoids absorb blue-green and blue light so they look yellow. Yes, carotenoids are also found in carrots and other yellow-orange vegetables. In the summer, carotenoids help absorb light and transfer the energy to chlorophyll to help with photosynthesis. Their contribution to tree color in the summer is to make a leaf brighter green.

In the fall, days get shorter and nights get cooler. Contrary to one popular belief that a frost is necessary, it is those two changes that trigger a built-in mechanism in the plant that slows down the replacement of chlorophyll. The leaf actually grows a membrane at the junction of the leaf and stem that slows the sap flow. As the sap flow is cut off, the amount of chlorophyll declines, and leaves with lots of carotenoids will begin to look yellow. These yellow compounds are more stable and linger much longer.

To complete the story, we need to talk about one more type of color that develops in leaves, anthocyanins. These chemicals form from a light-induced reaction between sugar and protein in the tree sap when the sugar concentration gets high. The sugar content in a leaf rises in the fall as the sap flow declines and some of the water evaporates. Add some sunny days, and the leaves make anthocyanins and turn shades of red and purple. Incidentally, you are already familiar with anthocyanins. They produce the red in an apple skin and the purple in a grape skin as those fruits ripen. The mechanism for the colors of ripening fruit is related to the leaf color mechanism. The sugar content of fruits increases as the fruit ripens, which, when combined with sunlight, produces the red and purple anthocyanins.

We now have all of the ingredients for the fall display. Depending on the species of tree, the amount of fall sunlight, the temperature, the amount of moisture and the acidity of the soil, we see the whole range of yellow, orange, red and purple colors. Add to these the rich browns of the leaves without the other pigments, and you have nature's fall extravaganza. Of course, mixed in with the trees and plants that lose their leaves in the fall are evergreens and some other plants that do things quite differently.

Add up just a few of the things that trees do for us –
oxygen, a shady spot to reflect, and a fall panorama – and it
sort of makes you want to go out and plant another one.

The Pot of Gold at the End of the Rainbow

Have you seen chromatic dispersion from reflective refraction of white-light illumination on the front surface of a cloud of suspended water droplets lately?

Better known as a rainbow, this charming sight is common after summer rain showers. It comes about when sunlight, always from behind you, strikes an area containing many small drops of water. The collection of water drops nearest to you acts exactly like billions of tiny prisms.

The white light from the Sun consists of a mixture of the full range of colors. As the light enters each drop it is bent (refracted) an amount determined by the color of each component, violet more than red. As the light reaches the back surface of the drop, it is reflected back toward the observer and bent some more. The emerging light is separated into bands of the colors red, orange, yellow, green, blue, indigo and violet. *Hint:* ROY G BIV is a trick to remembering the sequence of colors.

When the rainbow is especially bright, you can sometimes see the secondary rainbow outside the primary one. This much dimmer display has the colors reversed. The secondary rainbow results from internal reflection from both the front and back of the drops so that the light makes two roundtrips inside the drops before emerging. Theoretically, a tertiary rainbow is possible but rarely seen.

If you ever get the chance, look at a rainbow through polarized sunglasses. Try rotating the glasses. You may find that this enhances your ability to see the rainbow, or you may see a double rainbow, the second one inside the first without the colors reversed. Some of the light emerging from the drops is also polarized, and the glasses pass or block the polarized light, depending on how they are rotated.

Ever notice that you can never really get much closer to a rainbow if you drive toward it? Because the Sun must be directly behind you to see the rainbow, and it is the drops on the outside of the cloud that reflect and refract the light, the rainbow will stay in front of you until you drive into the cloud. At that point it will disappear. This is the reason for all of the stories of the magical things that will happen if you go to the ends or through the rainbow.

Since the ends of the rainbow are usually defined by the horizon, the best place to view one is often on a hilltop. If you are lucky enough to see a rainbow from an airplane, it can be a complete circle. You may see the shadow of your plane directly in the center of the circle.

A somewhat related phenomenon is the sun halo or the sundog. Particularly on a cold day when there are some high cirrus clouds, you may see a pronounced halo around the Sun as the ice crystals in these clouds reflect the sunlight. When the Sun is near the horizon in the morning or evening on one of these days, you may also see colored spots on either side of the Sun, called sundogs. It may be necessary to block the Sun itself with your fist or other objects to see them well.

Personally, I like physics. It's a shame, however, that it gets in the way of finding the pot of gold at the end of the rainbow. We can always hope that those laws of physics break down once in a while.

Connecting Bilirubin, Babies and Blue Light

A curious treatment for newborns, called phototherapy, is an interesting story that combines some practical wisdom with modern technology, accompanied by a mystery.

Newborn babies fairly often struggle with a condition called jaundice for the first few weeks of their lives. Jaundice results as the baby transitions from a shared circulatory system to their own. During this time, some of their red blood cells die. As their body breaks down these cells, one of the products is a yellow compound called bilirubin. The liver in an older child or adult takes up the bilirubin, attaches a sugar molecule to it to make it soluble in water, and excretes it into the bile. This "make bilirubin water soluble" step doesn't work very well in newborns for a few weeks, so they often develop the distinctive yellow coloration of jaundice.

Most of the time this condition, which is medically called neonatal hyperbilirubinemia, is harmless; but if the bilirubin levels get too high, it can begin to invade the brain tissue and cause serious damage. Because of the risk of brain damage, doctors closely monitor bilirubin levels in newborns and often implement treatment in an incubator surrounded by blue lights. This light treatment is called phototherapy.

The history of phototherapy is interesting. Many years ago, observant nurses in newborn nurseries noticed that jaundiced babies who were nearer the indirect light of windows improved noticeably. This led to clinical trials with artificial light sources. Blue light was shown to be the most effective. The reason for the effectiveness of blue light is predicted by the science of color. A yellow compound like bilirubin is yellow because it is absorbing most of the blue light coming to it. So, whatever the light is doing, you would expect the color of light that is absorbed the most to be the most effective.

Phototherapy was widely implemented as a simple, effective treatment for newborn jaundice. It seemed to be safe, but there was no explanation for why it worked. During the 1960's and 70's, the chemistry of bilirubin was studied extensively by several researchers. In particular, the photochemistry was examined. Photochemistry is the chemical reactions that a compound undergoes as it is exposed to light. Bilirubin was found to undergo a rich variety of photochemical reactions, but all of them were very slow. None of the photochemical reactions could explain the rapid improvement of an infant undergoing phototherapy.

In the late 1970's, researchers at the University of Nevada in Reno discovered a new photochemical reaction of bilirubin. This reaction had previously been missed because the product formed – nicknamed photobilirubin – is unstable and converts back to bilirubin. Photobilirubin turns out to be water soluble. As quickly as it is formed, it is taken up by the liver and excreted in the bile where it appears as ordinary bilirubin.

Phototherapy is still used, seems to be very effective, and is now a little less of a concern. Solving the chemical mystery has made everyone even more confident in the procedure.

Lightning Never Strikes Twice in the Same Place and Other Myths

Compared to other natural forces, lightning is actually a common event. Lightning strikes the Earth about 100 times per second, and there are between 20-40 million strikes per year in the US. Florida is the winner for the most lightning strikes in the US, but the central and southeast parts of the US are highly susceptible. How do we know this, you ask? Automated sensors, which have antennae and pulse counting electronics, are linked around the country in networks to collect data for the National Weather Service.

Lightning is not only very common it actually kills more people – 70-100 – each year than tornadoes and hurricanes. Curiously, it kills about four times as many men as women. This probably boils down to the fact that more men than women are outside during thunderstorms. We probably should not explore that too much further.

Like the rest of Mother Nature's arsenal, the power of lightning is awesome. An ordinary bolt of cloud-to-ground lightning has an internal temperature of 50,000 degrees Fahrenheit, much hotter than the surface of our Sun. The rapid heating of the air by this high temperature causes the sonic shockwave that we hear as thunder. The energy of a lightning bolt is estimated to be one or two billion joules. A billion joules would light a 100-watt bulb for 116 days. The fact that the bolt only lasts a few thousandths of a second makes it nearly impossible, unfortunately, to harness this energy.

The first myth we need to dispel is that if you are struck by lightning it is usually fatal. Actually, the mortality rate is about 20-30%. A peculiar characteristic of lightning is that it often stays on the outside of your body. When this characteristic is combined with the fact that a person is often hit with a side flash from the main strike, it means that a victim can sometimes be knocked down without serious injury. The frequency of structure strikes and the rather low mortality rate probably explains why there are so many lightning stories. Everybody seems to have one.

Lightning probably has more myths and half-truths associated with it than any other natural force. Most of us have heard that lightning always strikes the highest spot so:

- It's safer in the open than under a tree
- You should crouch if you are caught in the open in a storm.
- Don't hold metal objects like golf clubs, umbrellas, etc., if you are out in a storm.
- Get off the water if you are in a boat.

In fact, none of these things really matters very much. If you are outside during a thunderstorm, you are at risk. The electric charge that initiates a lightning strike begins miles up inside a thunder storm cloud. The charge rushes toward the ground as what is called a stepped leader. It is not affected by anything on the ground until it gets very close to the ground. When the stepped leader is 150 feet or so from the ground, another charged path will rise to meet it. If there is an object – such as a tree, a house, or a car – just below the stepped leader, the completed pathway will include the object in the lightning bolt. If there is nothing on the ground, the lightning will strike the ground, even if a tree is nearby. Lightning does not alter its path to strike any particular object unless the object is large compared to the miles-long path of the lightning bolt. A tall tower or building can affect the path of lightning, but not much else can.

So what does help if you are outside during a thunderstorm? The only thing that makes a big difference is to get inside a building or a car. Buildings or cars may still be struck by lightning, but again, the tendency of lightning to stay on the outside of a structure will make you much safer.

Unfortunately, wires, pipes, etc., can bring some of the lightning to the interior of a house.

Another myth is that lightning doesn't strike the same place twice. Very tall structures like towers and large buildings are struck by lightning many times. Their design must allow for very good grounding to prevent damage.

Whether it really matters or not, I am still going to stay away from trees and water if I am outside in a thunderstorm. That early training is hard to replace.

The Alien World of Deep-Sea Hydrothermal Vents

Imagine, if you can, a dark world, crushing pressure of thousands of feet of sea water, and temperatures that would boil sea water if it were on the surface. Although these conditions are close to the way we cook our seafood, a surprising variety of sea creatures call this formidable ecosystem home.

Hydrothermal vents are unusual columns of hot water that form from cracks in the sea floor. Think of them as big, undersea hot springs. They were only discovered in 1977 with a deep sea submersible near the Galapagos Islands. At that time, the vents were called "Black Smokers" because of the sulfur compounds dissolved in the hot water that precipitated out as the hot water cooled. The precipitating mineral sulfide particles formed a rising, black plume that looked like smoke. Since that discovery, there have been a number of other vents discovered in both the Atlantic and Pacific oceans. It seems a little embarrassing that we managed to miss these interesting features for so long. It makes you wonder what else is down there.

Although mildly interesting themselves, the startling discovery related to the vents was the diverse system of living organisms that surrounded them. Normally, at ocean depths of 2,000 feet or more, the seascape is mostly a wasteland. No sunlight penetrates to this depth, so there are no plant organisms and, consequently, none of the animals that would come to eat the plant organisms. Around the vents, scientists found colonies of giant tube worms, clams, shrimp, and a number of other animals. Tiny, single-celled, organisms called extremophiles apparently can live in the vents and use methane and hydrogen sulfide directly to manufacture proteins and the other chemicals of life in a process called chemosynthesis. This process is similar to the photosynthesis process used by plants, but instead of sunlight, they use thermal energy from the vent. The microscopic extremophiles are consumed by the tube worms and clams, and an entire ecosystem is established, all anchored by the energy and nutrients contained in the vents.

Extremophiles are living organisms that exist under conditions that, until recently, were thought to be incapable of sustaining life. Note that hydrogen sulfide, present in the vents, is similar to cyanide in its toxicity to most animals. The discovery of these kinds of organisms has completely changed the way scientists look at origin of life on Earth. Life was thought to be a very recent addition to Earth compared to its age. During the very early days of the Earth, it was a very inhospitable place with many volcanoes, meteor strikes, and no oxygen in the atmosphere. It was assumed that life came long after these early stages. Now, we can't be so sure. Now that we are beginning to understand single-cell extremophiles, it is quite possible they could have been here a very long time, beginning well before conditions settled into what we would normally consider life-sustaining.

This new view of the conditions necessary for life also changes our view of the possibility of life in our solar system outside Earth or, for that matter, in the universe. Instead of looking for ET, we had better be sure that we aren't missing a colony of giant tube worms somewhere. The ones here on Earth seem to be silent, but out there they might have something to say.

Bacteria, Viruses, Parasites, Fungi and Now Prions

It seems like every day there is a new plague to worry about. However, the science involved in an older scare – mad cow disease – is particularly interesting. This disease adds a completely different kind of infectious agent called a prion to Pandora's Box.

Infectious diseases in man are caused by one of four types of agents: fungi, parasites, bacteria or viruses. The Great Plague of the Middle Ages that killed millions of people was caused by a bacterium carried by rodent fleas. The Spanish Flu of 1918 that killed millions was caused by a virus. Malaria is caused by a mosquito-borne parasite.

Bacteria, fungi and parasites are all microscopic living organisms. When certain of these living organisms are introduced into an animal or man, they reproduce using the material of life, DNA, and cause damage of various sorts, which may lead to death. Infectious agent number four, viruses, lack the cellular machinery to live and reproduce on their own, but they also reproduce using DNA after they invade a living cell. A central dogma of biology has been that infectious agents must have DNA. That was about to change.

Beginning in the 1960's the agent causing a very rare, fatal brain disease called Creutzfeldt-Jakob Disease (CJD) was found to have completely different properties than any other infectious agent. In a breakthrough in 1982, a research group lead by Stanley B. Prusiner (Nobel Prize in Medicine, 1997) was able to isolate a specific protein that caused the disease. For the first time, a protein, a substance containing no DNA, was shown to cause disease. This type of protein was called a prion (pronounced pree-on). Prion was coined by combining the words proteinaceous and infectious. The staggering information was that a prion was not a poison but an actual infectious agent. Although the exact mechanism is still not understood, a prion can apparently bring about a structural change in a specific protein on the surface of a nerve cell. This structural change interferes with the connection of one nerve cell to another.

In fact, it was found that there is a whole family of brain diseases ranging from Scrapie in sheep and goats to bovine spongiform encephalopathy (BSE) in cows. BSE, notorious as mad cow disease, became a serious outbreak in the UK and parts of Europe in 1986. A clear link between this outbreak and the appearance of a new disease in humans called variant CJD was established. The likelihood that the prion from BSE was beginning to cause disease in humans developed into a worldwide scare that continues today.

Research quickly established that the BSE outbreak was a result of the practice of feeding UK and European cattle ground meat and bone products from other cattle. Also, existing beef processing methods in the area at the time allowed neurological tissue, including brain, to contaminate beef sold for human consumption. It was estimated that around 400,000 cattle infected with BSE entered the food supply in the 1980's in the UK and Europe. Because the latent period prior to symptoms can be years and, unlike all other infectious agents, prions are not destroyed by cooking, this information caused a near panic worldwide.

There is actually a certain amount of good news in all of this. Research in prions seems to indicate that, despite their name, they do not all cause disease. I won't be surprised if prions have a role in other processes, perhaps even the origin of life. On the mad cow disease front, not only have cattle feeding processes around the world been improved but testing of meat for BSE is now possible. Even better, there have only been about 160 cases of variant CJD in the world and the number of new cases seems to be dropping.

So, in spite of the fact that there are likely a lot of new vegetarians in the world, driving to the local hamburger joint is still a whole lot more dangerous than eating their fare.

Why Do People Get Old?

My impression of aging has always been that we have parts that wear out, not unlike machines. When the part that wears out is in one of our vital organs, we die. I also thought that our vital parts had a pretty definite lifetime of somewhere around 100 years. But this simple model of humans turns out to be quite wrong.

We humans have the characteristic that would be called redundancy in a machine. In a small plane, the engine has two magnetos. If one fails, the other can keep the engine running. Computers often have multiple processors, so they can continue to function if one fails. We have two kidneys and can live just fine on one. Even better, our vital organs are made up of tiny cells that constantly die and are replaced.

As a consequence of our redundancy, we do not fail like a simple machine. A graph of simple machine failure that plots rate of failure versus age would be a flat line that suddenly turns up as vital component lifetimes are reached. There would also be some early failures because of manufacturing defects. A graph of human rate of death versus age is more complicated.

Such a graph looks a little like a bath tub. Death rate, or mortality, begins fairly high with infants and small children because of the problems they are sometimes born with and some early childhood diseases. The rate declines as children approach about 5 years. This descending line on the graph represents infant mortality and is one side of the bath tub. From about 5 to 15 years the human mortality rate is fairly flat, like the bottom of the tub. From about 15 to 90 years, the mortality rate goes up rapidly. The rising line that makes the other side of the bathtub is what we call the mortality of aging.

Interestingly, the shape of this graph is very similar to the failure graph of complex electronic systems with redundancy. Without redundancy, the graph of a computer failure rate versus time is much like a simple machine. With redundancy, a computer shows a bath tub graph of failure rate, including aging, very similar to humans. Computers that have very high early failure rates also tend to have higher failure rates during aging. These data are similar to humans, suggesting that humans already have defects when they are born and redundancy copes as well as possible.

One more phase of the bath tub graph is also shared with redundant computers. Data show that around 95-100 years the mortality rate for humans flattens again. In other words, at 95 our mortality rate is high but about the same as when we are 110. Computers with redundancy show the same effect. This information suggests that there is not a fixed biological limit to longevity.

These data on aging suggest a model that describes us in old age. We are born with varying numbers of defects or damage. Redundancy takes care of those as well as possible and allows us to be tolerant of damage. A consequence of tolerance is that the damage can accumulate as we age. Eventually, the accumulated damage causes parts to fail as redundant mechanisms are overwhelmed.

More than just a curiosity, these data may point to things that are important for our longevity. Good pre-natal care is important for more than just reducing infant mortality. It also improves longevity in the early part of the aging process. Medical care and research into all conditions – such as hidden inflammation and infection, which may disable the redundancy mechanism and contribute to arthritis, heart disease, Alzheimer's disease, and cancer – may be very important for longevity. Research into growing new organs could compensate for accumulating damage. Avoiding behavior that directly or indirectly causes damage is important, too, of course.

I am not sure I feel any younger, but it's nice to know that Grandad, at 96, is out of the bathtub and into the flat part of the curve again.

Where Have All the Frogs Gone?

Beginning in the 1980's, observers all over the world began to notice an alarming decline in the numbers of amphibians. Amphibians are the class of fresh water animals that includes frogs, toads and salamanders. These interesting animals usually have an early aquatic stage called a tadpole that is equipped to breathe with fish-like gills. As the tadpole matures, it undergoes a metamorphosis into an adult equipped with lungs, so it spends less time in the water.

The decline in amphibian population was so severe that some species quickly became extinct. Scientists studying the dead and dying animals were unable to come up with an immediate, conclusive diagnosis. Amphibians in general have a rather sensitive skin, so many theories developed that the class was responding to air and water pollution, habitat destruction, climate change, or increased exposure to ultraviolet light because of a decline in the protective ozone layer. There was some data to support all of these possible reasons for amphibian population reductions, but none of them adequately explained the suddenness of the decline. The search for an explanation became a first class detective story.

One of the first clues in the search for an explanation was that amphibian mortality was not evenly distributed. It was worldwide, but it tended to occur in waves through a population. This tended to argue against an environmental threat or a generalized change in the habitat. Pollutants and habitat would be worse in some areas than others but would tend to be evenly distributed within a particular area.

In 1998, a common element was discovered in many amphibian mortality cases. The common element was a fungus called Batrachochytrium dendrobatidis. The fungus was found to cause a fatal disease in amphibians called Chytridiomycosis. I can't pronounce the name of the fungus or this newly discovered disease, but it appears to be devastating amphibians. Although not accepted by everyone, this explanation appears to be the most likely.

So, why did this disease appear suddenly? Should we care? What's the outlook? The "should we care" question is the easiest to answer. As we study the incredible diversity of plant and animal life – the biodiversity – on the earth we realize that this diversity is tightly coupled with who we are. Increasingly, we understand that seemingly small changes to one part of the bio-population can have large consequences to other plants and animals, not to mention reducing the richness of the experience of being human. The other questions about the new disease and its outlook don't have complete answers yet. Research is ongoing for ways to save the remaining amphibian populations. One theory is that the fungus has always been present in the fresh water environment, but environmental changes have made amphibians more susceptible. Another theory is that the fungus has changed recently, perhaps because of environmental changes, to become more of a threat. Whichever is correct, the fungus is found worldwide.

The outlook is grim but not hopeless. Some species of amphibians seem to be less susceptible to the disease than others. This offers some hope that future generations of susceptible species will develop some tolerance. Research will continue in ways to treat the disease and protect less affected populations.

The next time you hear a frog tuning up, wish him well. He's in a battle for his life.

Are Leeches and Maggots Coming to Medicine?

Can you guess what sort of a connection a wounded soldier in the Middle Ages has to the very latest, modern medicine?

Two things connect the soldier to today's medicine – leeches and maggots. Yes, really.

We'll talk about leeches first. If available, a barber would have treated the wounded soldier. A barber in the Middle Ages was a surgeon. Apparently, physicians considered surgery a dirty sort of business and didn't practice it. Besides amputations and various other extreme forms of surgery, the barber might have tried blood letting on the soldier (if he had any left.)

Since at least the time of the Greeks, there was a thought process that illness was due to an imbalance of certain fluids in the body. Blood letting was used to rebalance the fluids. Barbers sometimes heated a brass cup and inverted it over a small cut in the skin. As the cup cooled a partial vacuum was created inside and blood was withdrawn.

Barbers also used leeches for the same purpose. Leeches are little creatures that often live in swampy water. They can attach to the skin with a powerful suction mouth and quietly feast on your blood. The popular species for bloodletting was a native of Europe and became the medical leech.

Leeches continued to be used for blood letting to treat a wide variety of ailments up until the late eighteenth century. After that, it largely disappeared in most cultures because it rarely did any good. Current barbers have little connection to their surgical history, except for the barber pole outside. The alternate white/red pattern probably refers to the blood-stained tourniquets that were used, and the brass cap on top is probably the brass cup for blood letting. I'm not quite sure why barbers think they should remind their modern customers of this particular part of their history.

For a time, it looked like the leech chapter in medical history was closed. Then, about 25 years ago, medical leeches quietly reappeared in advanced surgery. When a surgeon performs the delicate task of reattaching a severed body part or performs certain types of plastic surgery, their success is sometimes limited by the accumulation of blood behind blood clots in the damaged tissue. The use of powerful anti-clotting drugs is risky because they can cause other problems. Along comes the lowly leech. It turns out that medical leeches can provide the perfect assistance in some cases. The leech can remove accumulated blood from a local area and release some natural anti-clotting compounds from their saliva into the same area – all without causing much additional damage. Medical leeches do not in general carry disease organisms anyway, but they can be grown in a laboratory to guarantee safety.

Now we come to the discussion of maggots. Our soldier almost undoubtedly has some of his wounds infested with maggots. Since maggots are fly larva, they are the hallmark of unsanitary conditions. However, maggots have one characteristic that makes them valuable to modern medicine. They only consume dead tissue. In the case of a wound complicated by gangrene or diabetes, maggots are experts at removing the dead tissue while leaving healthy tissue untouched. Turns out they are much better surgeons in these cases than the medically trained ones. The use of maggots is not routine, but in certain cases they can save a limb from amputation. They also can be bred in laboratory conditions to prevent contamination from disease-causing organisms. Yes, both leeches and maggots are approved for use by the US Food and Drug Administration.

If, like me, you still go to barber shops instead of the newer hair salons, you might want to make it clear that you are feeling great, and you just want your hair cut.

Can You Hear in Colors or Taste Shapes?

Although you may not be able to, and I certainly can't, a few people can. These people have a rare abnormality called synesthesia. It was first described and named in 1880 by Francis Galton. Since that time, the condition has been mostly ignored. It was generally dismissed by the scientific world as people with vivid imaginations, suppressed memories, or even people who were deliberately trying to mislead.

Synesthesia comes in several variations, but all appear to be a sort of mis-wiring of one or more senses to the wrong part of the brain. People are generally born with the condition, but sometimes a brain injury or a drug can induce the effect.

In one variation, a synesthete "hears" a particular musical note as a particular color. It is a little hard for us to imagine, but these people hear normally and also see colors within their field of vision linked to the sounds. When they hear a C sharp musical note, they may see a shade of green. The connection of particular musical notes to particular colors is consistent for an individual but varies from one synesthete to another. You can imagine that this condition is not a handicap but might add a special richness to music. We also wonder how many composers had this condition.

Another variation is that a printed number or letter is always perceived in color. The printed number 5, even though it is printed in black, may appear red to the synesthete. Once again, the perception of color is very consistent and may not include all numbers or all letters. The font (the type of print) that the number or letter appears in may influence how bright the color is. If a synesthete sees an E as green and a red letter E is shown to the person, he or she feels uncomfortable but recognizes the real color of the letter after a short delay. Some synesthetes can change the color of a letter by adding a line to it. For example adding one line to a P to make it an R might change its color from orange to yellow.

An additional variation is people who perceive a taste in their mouth when they touch or see particular shapes. It makes us wonder how many of our metaphors, such as sharp cheese or loud shirt, arise because we all have some synesthesia.

Synesthesia began to be taken much more seriously in the last couple of decades when some clever tests were devised for the condition. One such test could be called a "pop-out" test. Ordinary folks would be very slow to recognize a pattern composed of printed black 2's in a background of black 5's. A synesthete who sees numbers in color would instantly see the pattern if the 2's appeared to be red and the 5's appeared to be green.

A higher tech approach to the study of synesthesia is called Functional Magnetic Resonance Imaging (fMRI). This medical imaging technique allows a researcher to see which part of the brain is active when a subject is concentrating on a task. Research using this tool clearly shows that people with synesthesia use different parts of their brain from normal people when they are calling on the affected sense.

Because the condition has been clearly identified as real, we can expect more details to appear as it is studied further. As abnormalities go, this particular one seems like it might be more of a gift than an affliction.

Tales of 100-Foot, Ship-Devouring Walls of Water at Sea

Imagine you are crossing the ocean in a ship. You feel secure in the knowledge that you are aboard a large, well-designed ship with an experienced captain. Sea conditions are somewhat stormy with 10-foot waves and an occasional 20-foot one. The ship is pitching but easily handles the 20-foot waves.

Suddenly, you see a single wave approaching as tall as a 12 story building – a wall of water over a hundred feet high. It breaks over the ship, knocking out the windows on the bridge and ripping deck equipment from mounting hardware. Water enters the engine room. The engines and electrical generators stall. The ship is dead in the water. Its crew and passengers sustain minor injuries. Luckily, the ship was headed almost directly into the wave, or it might be on its way to the bottom, another unexplained ship disappearance.

The incident described above actually happened to two ships – the Caledonian Star and the Bremen – in the south Atlantic within days of each other. Until very recently, these stories were the stuff of maritime lore and an occasional movie plot. Experts in the field of ocean waves said their models didn't allow for such waves. Captains who survived such rogue waves sometimes minimized the event to avoid being placed in the same category as those who have encountered UFO's.

Unexplained ship disappearances have plagued shipping for centuries. Today, they are not common, but they continue to happen at the rate of one ship per week. Many of these losses can be explained by poorly maintained ships or careless navigation, but not all of them.

In 1995, things changed suddenly. The Draupner oil platform in the North Sea made history with accurate measurements of a wave of nearly 100 feet high on January 1, 1995. The platform survived with some damage, but the measurement created a stir in the scientific community. A flurry of satellite and deep sea buoy studies began to create a scary picture. Not only do 100-foot rogue waves exist, but they are not uncommon and are found all over the world. Although the chances of a ship encountering one are still fairly low, clearly many of the myths in history are based on fact.

Rogue waves are still a hot topic for research. No single theory explains their formation. Although some areas are more prone than others for extreme wave formation, the waves can form anywhere there is deep water. Such waves have been detected on the US Great Lakes and inland seas. A rogue wave is a surface phenomenon that typically does not travel great distances and isn't usually a threat to land-based structures. These waves are completely different from tsunamis, which are generated by undersea earthquakes. The tsunami is almost undetectable in deep water and becomes a major threat only as it travels (often great distances) to shallow water and land.

The lack of real predictability and the poor feasibility of a warning system create a dilemma for the global shipping industry. Ships are currently designed for what was thought to be worst case waves. It is physically and economically impractical to design ships for a wave that is DOUBLE what was thought to be the worst case. Until a great deal more is known about rogue waves, they will remain another powerful natural force from which we have little protection.

I would very much like to see one of these waves – but only from an airplane.

Do You Know Your Left Brain from Your Right Brain?

It's interesting how some results of studies in neurophysiology have so quickly woven their way into our popular culture. How many times do you hear people refer to right- or left-brain thinking? How many advertisements have you seen for books or techniques to "unleash" your intuitive, creative right brain? How much of this popular left/right brain stuff has a basis in fact?

Fact: The higher part of the human brain, the cerebral cortex, consists of two large lobes on the left and right side of the head, connected by tissue called the corpus calossum.

Fact: The left side of the brain generally controls movement on the right side of the body, and the right eye sends its visual information to the left side of the brain.

Fact: If the left and right cortex are surgically divided by cutting the corpus calossum, the patient can function normally but has some unusual characteristics. Special tests on these patients provided insight into each side of their brains separately. In general, the left side of their brains had sophisticated speech capability and a rational, intellectual style. The right was generally inarticulate but blessed with special spatial abilities.

Fact: Brain scans on normal people performing various tasks support some left/right specialization but also show complex interactions.

The characterization of the split-brain patients mentioned above and the correlation to normal brains has been challenged. The reason that these patients had their brains surgically divided was that they were afflicted with very severe epilepsy. The surgical procedure was found to interrupt the electrical storm that moved across their brain and generated seizures. They had, in general, suffered from seizures their whole lives. The brain damage that caused the epilepsy in the first place could have altered the localization of brain functions.

Even the trend toward speech localization in the left side of the brain is complicated by left-handed people. Although 95% of right-handed people have speech localized in the left brain, about 18% of left-handed people have speech localized in the right brain.

Based on a liberal interpretation of these facts, the marketing began. Countless books, articles, and seminars promise to "put you in touch with the creative right side of your brain." You have probably seen the young woman/old woman picture, the revolving dancer, and several others that change as you look at them. These have been touted as evidence of the divided brain. Learning styles and talent have been attributed to the dominance of one side of the brain.

Unfortunately, much of the popular information about left/right brain is either a gross simplification or completely in error. Interestingly, the misinformation seems to be mostly harmless. Many of the techniques described to make a person more creative or more artistic actually do work. Even though the tricky pictures that change with viewing can be explained with ambiguous visual cues built into the pictures and not a divided brain, they are still fun. Some years ago I read a popular book called *Drawing on the Right Side of the Brain* by Betty Edwards. I now think the divided brain premise in the book is of no importance, but it is still an excellent book if you want to learn how to draw. It is, in fact, the only book I have ever found that provides simple exercises that can teach anyone to draw in a short time.

For now, if someone accuses you of only having half a brain, you can either say "yes, but I have maintained the highly intellectual, verbal left brain," or you can gesture toward yourself and grunt "da Vinci." You could also just use the very primitive part of your brain and throw something at the accuser.

The Tiny, Often Misunderstood Creature with the Mighty Itch

At a recent music lesson, my teacher was constantly scratching his ankles. When he finally removed his shoes and socks he, was covered with dozens of red welts – the dreaded chigger bites. As is often the case, he asked me what a chigger is and why they bite. I rattled off all of the conventional wisdom that I had collected from many years and many bites, telling him a chigger is a tiny, blood sucking insect that burrows into the skin. Later, I decided that maybe I should check on how factual my knowledge really was. In fact, the only thing I was right about was that they are tiny.

A chigger is the larval stage (between the egg and the nymph) of the harvest mite. Mites are in the class Arachnid, so they are related to spiders, not insects. The chigger is nearly microscopic and orange in color, so they are invisible on the skin. They feed on all kinds of animals, including humans. Harvest mites are very widespread in temperate zones of the world.

The simple rule I learned to distinguish a spider from an insect was to count the legs. Spiders have eight, and insects have six. This still works fairly often, even though spiders actually have 6 pairs of appendages. Two pairs have been adapted for eating and defense. A better rule might be that spiders don't have wings or antennae. With mites, the legs are even more complicated. The chigger (the larval stage) actually has six legs. When it becomes an adult, it has eight.

Rather than being blood suckers – like their relative, the tick – chiggers eat skin cells. They attach themselves to the skin, inject an enzyme, and suck up the pre-digested skin cells. This initial process is painless. After receiving their fill, they drop off, mature to a nymph, and become an adult that feeds strictly on plant matter. The tiny parasite is long gone before the itch starts. Therefore, all of the home remedies that are designed to remove the chigger from the welt are too late. The hole they leave behind becomes highly irritated, and the surrounding tissue swells in response.

There are many remedies for the itch. The best may be one of the over-the-counter anti-itch creams containing hydrocortisone.

Another common story about chiggers is that they lay their eggs in the bite. Since chiggers are babies, they can't lay eggs. The adults lay eggs on plant leaves and have no interest in animals.

Avoiding chiggers is not easy unless you use insect repellant. They are found in forests and grassy areas. They avoid sunlight and humidity, so closely cropped, watered grass in direct sunlight is less likely to yield the pest. They are more likely to be found in untended grassy areas that are shaded. The first frost of the year generally eliminates them until the next summer.

Chiggers transfer to your skin and clothing as you move through grass and brush. They seem to have a tendency to seek areas to feed where clothing is tighter, such as at the top of your socks or around your waist. They can be washed off by showering immediately after exposure. My mother told a story of an outing when my brother was a baby. She was concerned that he had been exposed to chiggers, so she bathed him. Unfortunately the chiggers apparently just moved to the top of his head.

A chigger bite always seems to cause maximum misery. The welts can itch for a week or more. Scratching them makes them worse. If there is any good news about chiggers, it is that they don't appear to be significant disease carriers.

The Iceman Is a Case History for Both Prehistoric Archeology and Modern Conflicts

The Iceman is not a new story, but it is in the news again. If you haven't followed the story, this is a quick summary. In 1991, a partly frozen, mummified body was found in the Alps on the border between Austria and Italy. The German hikers and Austrian officials that were initially involved assumed the body was an unfortunate – but modern – mountaineer. Officials used a jackhammer (and other non-scientific methods) to remove the body from the ice and moved it to an Austrian morgue.

It quickly became clear that the body was actually very old – 3300 B.C. to be exact. It was also determined that the body was discovered a few meters inside the border with Italy, so Italian officials claimed it, thus beginning the first of several legal squabbles. Over the years, not only did the hikers who found the body claim a finder's fee (due to the high archeological value), but some others who probably didn't find the body also laid claim to a fee. Since I find the science much more interesting than the modern conflicts in court, I'll move on.

The Iceman now rests at South Tyrol Museum of Archaeology in Bolzano, northern Italy. He has been scientifically studied with modern crime scene and archeological methods in fascinating detail. He was 45 years old. Isotope analysis of his tooth enamel pinpointed the local areas where he lived his life. A piece of flint arrowhead and artery damage was found in his shoulder. A deep cut that occurred a couple of days before death was found in his palm, and a major injury was found to his head. The time in which he lived is often called the Copper Age, and he carried a well-crafted, copper-bladed axe. Tissue samples indicated high enough levels of copper that he was probably directly involved in the cooperative smelting of copper and the manufacture of copper objects.

His last two meals consisted of chamois (a goat-like animal which was native to the Alps) and red deer meat with some refined grain that could have been bread. Both meals were accompanied with some roots and fruits. Traces of pollen in his meals even showed at what altitude he was when he ate.

The most recent data in the scientific news concerns an interesting attempt to trace the Iceman's movements just prior to his death. Using traces of several species of moss and other contents of his gut, scientists were able to show that he traveled to a bog at low altitude to obtain some mildly antiseptic bog moss, probably to dress the palm wound and perhaps to wrap some of his food. He unintentionally ingested some of the moss as he ate his meals. Very likely the total trip length over about three days was about 37 miles.

We can only speculate about the conflict that ultimately killed him. Either the shoulder or head wounds could have been the cause of death. The head wound could also have been caused by a fall as a result of the arrow wound. DNA analysis of blood found on his clothing and weapons came from four different people. Clearly, he was involved in a heated battle, but a blood stain on the back of his clothing could be from a comrade that he carried.

Unfortunately, the Iceman was the only body preserved in the ice. Because he is so well-preserved, some unusual circumstances must have covered him with snow soon after death and kept him covered in ice for the five thousand or so years afterward.

What Really Happened to the Dinosaurs?

The scientific method is a proven means to approach the truth, but it can be slow with numerous, sometimes contentious, technical arguments. This makes it fun for us to follow the discussions.

Most scientists have agreed for a while now that something fairly catastrophic ended the era of the big dinosaurs and many other plant and animal species about 65 million years ago. The event marked the end of the Cretaceous geologic period and is one of several mass extinctions in Earth's history. Evidence of mass extinctions is obtained by studying deposited layers of sedimentary rocks that make up the geologic record. For example, 65 million years ago, fossils from many species suddenly disappeared in layers deposited after that.

In 1980, a team led by Luis and Walter Alvarez found a definite layer in sedimentary rocks at many sites all over the world that contained very high levels of the element iridium. The element is rare in the Earth's crust but common in large meteorites and asteroids. The layer dated to 65 million years ago. Estimating the total amount of iridium in the layer, the Alvarez team proposed that it could have been formed by the impact of an asteroid around six miles in diameter. Such an impact would have been the equivalent of two million thermonuclear bombs and would have generated a huge dust cloud. As the dust settled over many years, the layer that the Alvarez team found would have formed. The global climate change created by the dust certainly could have caused mass extinctions.

At first, the theory was ignored because the impact crater should have been on the order of 155 miles in diameter, and no such large crater had ever been found. Everything changed in 1990 when exactly such a crater was discovered on the coast of Yucatan, Mexico, partly under water. The crater, called Chicxulub, was named for the town that sits nearly in the center. The scientific data gathered on the crater shows that it is about 112 miles in diameter and is dated to 65 million years ago. In addition to a huge global dust cloud, there is evidence that the impact created gigantic tsunamis that went far inland in the surrounding land masses.

A number of scientists were satisfied that the mystery was solved. Some others were not.

Another event on the other side of the world in what is now India was happening at about the same time. A huge volcano was releasing floods of basaltic lava that formed what are called the Deccan Traps in central and southern India. These lava flows cover an area of at least 200,000 square miles and are almost 6,000 feet thick. These flows are one of the largest volcanic events known. This monumental volcanic eruption was also spewing gases into the atmosphere and would have had a massive effect on the climate. The climate change could have caused the mass extinction.

The evidence is clear that both events happened. Neither side of the debate claims that the other event is not real. Both sides claim that the other event could have contributed to the extinction, but both argue that their event had the major effect and happened first. The argument is not an easy one to resolve. The key is determining exactly when each event happened and how long the devastating effects lasted. The volcano was the slower process and could have been the major driving force. Unfortunately, the dating process for the actual extinctions and the climate changing events themselves have not, so far, been accurate enough to answer the question.

The answer could also be more like the perfect storm of two combined events. The debate continues, but that's what makes it interesting.

The Fireball in the Sky of 1908

Tunguska has been in the news again lately. If you haven't heard the story, Tunguska is a remote area in Siberia. On July 13, 1908, around 7:00 a.m., witnesses saw an intense fireball that seemed to split the sky. This was followed by several explosions that knocked people off their feet and broke windows up to 100 miles away. Seismic stations across Eurasia registered the event. Although it had not been invented yet, some say the event might have been equivalent to a 5 on the Richter Scale. Barometric pressure instruments as far away as Great Britain detected the atmospheric shock wave.

The question has always been: What happened at Tunguska?

It was assumed at the time, with little scientific interest, that a large meteor had struck the ground. Because the area was so remote, and Russia was in such disarray from World War I and the Russian Revolution, it was 1921 before there was a documented expedition to the area. At that time, a group was sponsored by the Soviet government to see if a large iron meteorite could be recovered for the metal.

What the expedition found was surprising. Instead of a crater, they found an area about 30 miles across of blown-down trees. It is estimated that 80 million trees over 830 square miles were affected. In roughly the center of the area, trees were upright but stripped of bark and limbs and charred. The results were exactly what would be expected from a mammoth explosion several miles above the surface. Estimates of the size of the explosion have varied over the years, but around 10-15 megatons of TNT (or 1000 times the Hiroshima A-bomb blast) is a common estimate.

Many theories have developed over the years as to *what* exploded. Some wonderfully bizarre notions include a "natural H-bomb," a small black hole, a blob of anti-matter and, of course, a UFO. A more reasonable theory was either a small comet or asteroid.

Real data was slow in coming because of the veil of secrecy surrounding everything in the Soviet Union. In more recent years – unfortunately, long after the event – the best data has been available. Analysis of tree resin and soil samples from the area indicates small particles that are high in content of the element iridium. This is a characteristic of stony meteorites rather than comets.

The most recent news comes from a supercomputer simulation undertaken at Sandia National Labs. Scientists there found that an object much smaller than previously estimated can explode when it builds up pressure ahead of it as it penetrates deeper into the atmosphere. When it explodes, a fireball can move downward as a column or jet faster than the speed of sound. This can produce great devastation directly below, even when the overall explosion is 3-5 megatons.

We will probably never know much more about the Tunguska event, but it makes me think that we need to study the detection and possible deflection of fairly small objects as they approach us.

Water That Runs Uphill and Other Mysteries

In scores of places around the world and several here in North Carolina, there are roads where spilled water runs uphill. If you drop a ball or any round object, it rolls uphill. Although not recommended, if you put your car in neutral it will roll uphill at increasing speed. If you walk up the road, it is unusually easy.

These places generally have similar names like Magnetic Hill, Mystery Hill, Spook Hill, Haunted Hill, or Gravity Hill, or the corresponding name in the native language.

Explanations for these strange places run from fairly reasonable scientific ones to the completely supernatural. Common explanations based on science include magnetic deposits or magnetic anomalies. The magnetic theory falls apart quickly because non-magnetic materials, like water, are just as affected as magnetic ones. Gravitational anomalies are a little harder to eliminate. Very unusual distributions of mass in the surrounding terrain might alter the local gravitational field. The gravitational possibility is, however, eliminated by making sensitive weight measurements along the road. So, what's left?

Many of us have discovered over time that, more often than not, the simplest explanation for something is correct. Country doctors discovered this long ago, although some more modern doctors sometimes forget. Even fixing the most complex electrical device starts with checking that it is plugged in.

The fellow who receives the credit for this wisdom of grasping the simplest explanation was a Franciscan Friar named William of Occam, who lived in England in the 14th century. In his writing, the most common statement of his thought is (translated from Latin), "Plurality ought never be posited without necessity." This later became known as

Occam's Razor. The reference to a razor was the analogy of "shaving away what is not necessary." This thought was so powerful that it has survived to the current day, still with his name on it. It is also possible that it has contributed to the scientific method, which depends on observations and experiments to eliminate alternate explanations.

Back to the mystery hills of the world. The simplest possible explanation for a road that looks uphill but acts in all other ways as if it is downhill is, of course, that the roads *are* downhill, and our eyes are deceived. Most of these places have been tested and are, indeed, optical illusions. In all cases, the terrain around the road confuses the viewer into thinking that it is uphill.

Mystery houses around the world are generally the same effect. The floor of a room in the house looks level but actually tilts, or vice versa.

Locations such as these in North Carolina are Mystery Hill in Boone, Old NC-21 in Thurmond, Richfield Road in Richfield, and the road between Maxton and Laurinburg.

There are a couple of places in Hawaii with Mystery Hills. If funds were available, I would offer to check them out.

Endangered Species: Desperate Cause or Folly?

If you consult the US official list of endangered or threatened animals, you find over 1,200 species. In almost every case on this list, there is some activity to stabilize the population of the endangered animals. In a number of cases, the activity is nothing short of herculean.

A case in point is the Whooping Crane. This magnificent bird stands nearly five feet tall with a wingspan of seven and a half feet. This large migratory bird came very close to extinction, with only a couple of dozen birds reported in 1941. Loss of wetland habitat from its breeding ground in Canada all the way to its wintering grounds along the Gulf Coast is the major factor in the Whooping Crane decline.

Beginning in 1967 when the crane was declared endangered, attempts have been made to establish breeding populations in the wild. Whooping Crane eggs were transplanted into Sandhill Crane nests. Although the chicks were successfully raised and learned to migrate, they imprinted on the foster parents and refused to mate with other Whooping Cranes. The project was discontinued as a failure.

Another attempt was made to establish a non-migratory population in Florida. Although still continuing, this experiment has demonstrated that Whooping Cranes must learn most of their behavior from adults. As a consequence, the Florida population has suffered decline from predators, such as alligators and bobcats. The birds didn't have the necessary skills to recognize and protect themselves from predators. The birds are also not reproducing in this environment.

The most ambitious project of them all involves birds that are isolation-reared in the north and trained in basic skills by handlers using hand puppets and costumes to imprint the chicks on Whooping Crane appearance. When they are ready for flight, costumed handlers flap their wings and run along the ground. When the chicks are airborne they are trained to follow ultra-light aircraft with costumed pilots. At the appropriate time, a group of volunteers led a migration along a new flyway established just east of the Mississippi River from Wisconsin to Florida using ultra-light aircraft. The journey is a difficult one because the aircraft are suitable for fair weather flying only, and they must stop at established sites along the route for refueling of the craft and pilots. The sites must be chosen to provide safe overnight roosting for the birds with costumed handlers waiting. The jury is still out on this migratory population, but some birds have successfully migrated without the aircraft, and there has been some successful breeding. The total US population of Whooping Cranes numbers a few hundred. This recovery is nothing short of miraculous, but the outlook for this great bird is still not good.

The story for many other endangered species, great and small, is similar. Mankind has removed specific elements of the required habitat. In some cases, the required habitat can be recreated or restored. In many others, mankind's requirement for energy and other resources has superseded wildlife's needs. In these instances, the only way to save the species is by propping it up with heroic efforts that are probably not sustainable for the long term.

The corporate mentality would say the unsustainable support of endangered species is economic nonsense. The activist would say that preventing the disappearance of multiple species is priceless and worth any effort that it takes.

The dilemma won't be solved soon. If there is a bright side to all of this, it is that there are so many people who feel that responsibility for creatures and their planet is one of the things that makes us human.

Surprisingly, Those Puffy Clouds are Infected

Every so often I run across research that has taken a bizarre turn. The following research is a great example.

Workers in meteorology, the science of weather, have proposed for a long time that tiny particles suspended in the air trigger rain drops to form in a cloud that is saturated with water vapor. The drop usually begins as an ice crystal that grows around a tiny particle. The crystal grows until it is heavy enough to fall out of the cloud. The result is precipitation in the form of rain, snow or hail, depending on temperature and other conditions. Hail is formed when strong updrafts in a towering thunderstorm cloud carry ice crystals to high altitudes, where they grow into a hailstone. The hailstone finally drops when it gets heavy enough to counteract the strong updraft; the greater the updraft, the bigger the hailstone.

The role of these particles, called Ice Nucleators (IN), in precipitation is substantiated by cloud seeding experiments. In this process, a load of small particles is dumped into a cloud from an airplane to produce rain. Even though the process works, it is usually less than practical because it still requires a cloud and depends on an expensive delivery system.

Recent research has shown that bacteria may be a very important source of the particles that produce precipitation. Rather than causing disease, these bacteria have evolved the ability to make it rain (or snow or hail). They apparently distribute themselves around the world on the winds. When they trigger precipitation, they fall to earth and live in soil and on the surface of plant seedlings. When conditions are right, they are picked up again by the wind and are ready to form ice crystals.

The IN bacteria are a few specific species that are found everywhere in the world and appear to be a major factor in precipitation. This research would be just a curiosity if it were not for the possibilities it raises. Perhaps the biology of IN bacteria is an integral part of the worldwide precipitation cycle. Precipitation patterns, of course, have a huge impact on the cultural and economic maps of the world. Since modeling of climate is still very inaccurate, perhaps the biological characteristics of IN bacteria are a missing element in current climate models. Further knowledge of IN bacteria might allow improvements to climate models. Better climate models mean better prediction of weather.

The power of the scientific method has always increased from cross-disciplinary interaction. I don't think, however, that meteorologists expected to need bacteriologists to work on the next climate model.

Perhaps IN bacteria could be the future of man's ancient desire to actually do something about the weather. Better understanding of the tiny creatures in clouds might allow us to eventually control when and where it rains and snows, even though I am not convinced yet that weather control is a good idea.

Rain dancing could also be effective. Clearly, if you do enough dancing, you will eventually be dancing when it is raining.

Topics in the Methods and Philosophy of Science

The Scientific Method – Or, Science Versus the Gremlins

If someone were to suggest that you use the scientific method to solve a problem, you might have images of laboratories and scientists. Perhaps you studied the scientific method in school. Your response might be, "I don't need anything very scientific. If my problem is a puddle of water under the kitchen sink, I need a wrench (or maybe a plumber, depending on my skills) not an experiment, a hypothesis or a theory."

Actually, the scientific method has three important characteristics. (1) It is probably somewhat misnamed. The scientific method is really a collection of principles, methods and an overall outlook, none of which are reserved for scientists. (2) This collective seems to include the most powerful tools ever used by mankind and clearly took us from the dogma-filled Dark Ages to men and women traveling in space. (3) It is not a magical, step-by-step formula.

If there is no magic and no single method of science, what explains the progress from wooden wheel to space shuttle in a mere 4,000 years? What is it about science that works?

The scientific method that is often taught is: observe, generate a preliminary explanation (hypothesis) for what you observe, test the hypothesis, and then develop a theory based on the results of testing the hypothesis. Theories then continue to be tested until they are proven wrong. Once in a while, a theory survives long enough and is fundamental enough to become a law of science. The success of this approach has moved science and technology to where it is today.

Scientists and others who are studying something don't always follow all of these steps. A common approach is for a scientist to start with an idea or a problem and study what has already been done in the technical literature. Communication and information technologies now play an exceedingly important role in science because they often show that an idea or problem has already been studied or solved. If an idea is unique, the scientist may work to fill gaps in what is known. Rarely does the work lead directly to a theory. Most work remains a hypothesis with some supporting data, and the scientist moves on to another idea or another problem.

The power of the scientific method is in the principles and the outlook that have developed with the methods. Some would argue that it is impossible for a person to explore their own idea without influencing the outcome. Scientists, like any of us, have strong beliefs, preconceived ideas and even prejudices. Their ideas can certainly be wrong. Ethical principles, peer review and the certainty that any important work will be repeated and examined by others have prevented most fraud and most serious mistakes. Because it is understood how easily the observer could unintentionally influence the outcome of a study, techniques like double-blind studies, statistical analysis, technical auditing and analytical protocols have evolved to separate the observer from the results. The scientific method has built-in self-checks that have withstood the test of time.

In the Dark Ages, a scientist caught challenging dogma was often imprisoned or, worse, could not teach science. Today, there may be no shortage of cherished, vigorously defended theories in science, but there is always someone out there trying to shoot the theory full of holes. Today, teaching and studying science are respected professions. The scientific method is alive, and the history of science and technology from the oxen cart to the Mars Rover shows that it works.

Back to the puddle under the kitchen sink. You and the plumber are probably much more scientific than you might think. If you check to see if the leak continues with the hot and cold water shut off, and then suspect the drain trap, you just used the scientific method. One of its key characteristics is that it feels natural to any practical person. Of course, if it still leaks after you replace the drain trap, it has to be the gremlins, right?

The Nature of Research

Research in all areas of science gets quite a bit of press coverage, particularly research projects that succeed or fail in a spectacular way. What about the failures? Are the failures a good measure of the quality of US science and technology? Are we slipping? Are the failures a result of the waning interest of US students in science and math or our ability to teach those subjects? What do failures tell us?

A fairly recent failure (2004) is the returning capsule of the Genesis mission. This mission collected samples of particles from the Sun, the so-called solar wind, and returned them to earth. Instead of floating down on a parachute and being gently snagged by a helicopter, the capsule plowed into the desert at over a hundred miles per hour. The parachutes didn't operate. Another fairly recent failure is the crash of the Mars Climate Orbiter, which was blamed on English/metric unit confusion. The Mars Polar Lander has still not been located on the surface of Mars. The failure mechanism is not known. Of course, two of the space shuttle missions have ended with loss of the astronauts. Often these disappointing, even tragic, results are characterized as unacceptable sloppiness.

Realistically, all of these spectacular events have something in common. They are research projects. The exploration of space can be directly placed in the research category. The nature of research builds in the element of failure because research probes at the edge of what is known. In fact, it may not be fair to ever call the result of a research project a failure. There is no way to carry out exploration and research and be completely assured of the result. If positive results are assured, then the project is not research. If transportation is safe, it's a bus not a space shuttle or a moon rocket. Astronauts are not passengers. They are explorers. Arguments can be made that exploration should be done by robots instead of humans, but humans have an inherent need to explore in spite of risk. Research into new cancer treatments show that they either help, or they don't. We expect cancer research to continue even if many results are negative.

If the outcome of a project is certain, then the research is either not research at all, or it is not aggressive research. The best measure of research is probably not success but failure. This is a concept that high-tech companies have struggled with for years. Management structures reward positive results, but good research will produce negative results a significant part of the time. The percentage of negative results is probably the best measure of the aggressiveness of the research. Aggressive research, often important research, will produce many negative results. No one advocates sloppiness, but if a procedure is complicated enough and not completely understood, mistakes will happen. These are also part of research. The strange nature of research means that major discoveries sometimes result from the mistakes.

The best researchers are flexible, alert and stubborn. The Genesis mission researchers will salvage everything possible from the crashed capsule. They may still be able to answer some of their questions about the origin of the universe from this spectacular "failure."

Putting Age in Perspective

I've been feeling a little old lately, so maybe it's time for me to put things in perspective. Talking to Grandad is always a good way. "You're not old," our 94-year-old resident elder says. "You're only around 60. That's the prime of life. You have only been earning a living for a little over 30 years. I've been retired longer than that! Besides, some days, I don't even feel old."

Feeling better already, I decide to reflect on the age of the universe and really get things in perspective.

Cosmology is the study of the structure and origin of the universe and includes studies of its age. As you might expect, cosmology is not an easy science to pursue. Mankind is a very recent addition to a very old universe, so cosmology depends on data collected by astronomers only after the telescope was invented in 1609.

The current estimate of the age of the universe is 13.7 billion years, so cosmologists are forced to make estimates about something that happened billions of years ago based on data gathered only in the last couple of hundred years. The good news is that, in the last few years, astronomy has taken giant leaps forward. The Hubble Space Telescope, combined with ground-based telescopes using sophisticated adaptive optics, has provided a wealth of excellent data in the last 20 years.

Even with good astronomical data, how do you go about estimating something that happened billions of years ago? There are three primary methods.

Method one involves a theory that cosmologists and astronomers have been working with for a long time, called the Big Bang. This theory is badly named because it doesn't refer to a normal explosion. The theory basically states that the universe was very hot and very dense in the beginning, and space itself began to expand at the birth of the universe.

This expansion of space continues today and causes all major astronomical objects to move away from each other.

This theory received a major boost in the 1930's when Edwin Hubble (honored by the naming of the space telescope after him) actually measured the speeds of numerous distant stars and showed a clear relationship between distance and speed. With today's best astronomical data, an extrapolation of current, distant star positions back to the time of the Big Bang gives a value of about 13.7 billion years. Also, careful measurements of microwave radiation coming from deep in the universe from all directions could be left over from the cooling that occurred during the Big Bang expansion. These measurements also point to a similar age. Both of these measurements, however, depend on the Big Bang theory being correct.

Method two involves estimating the ages of clusters of stars in our Milky Way Galaxy called globular clusters. These clusters are the oldest stars in our galaxy. Astronomers have determined this by studying stars of various ages and matching their characteristics with their ages. They have determined that most stars follow a well-defined sequence of changes called the main sequence. Measurements of the oldest stars in these clusters tend to point to the same 13 billion years.

Method three involves measurements of the element thorium in old stars in our galaxy. Thorium and other elements can be measured by analyzing the light from a star. Thorium is a radioactive element that decays to other elements. This decay proceeds with a constant rate such that half of the original thorium is gone after 14 billion years. This difficult measurement has been done for a few stars and also produces a similar age.

These values for the age of the universe will continue to be challenged and refined as the scientific method requires, but I'm feeling younger by the minute.

Technical Generalists Are a Vanishing Breed

A year or so ago I met a classic, technically knowledgeable generalist – in a rather unlikely place. He has a one-man business listed as an automobile tire mounting, balancing and wheel alignment shop. I had a problem with a new set of tires from a big tire shop. After multiple re-balancing, one wheel was still causing a rattle of one's teeth at 60 miles per hour. Like a specialist, this shop owner quickly determined that one of the new tires was not quite round and needed to be replaced. He gave me the data from his balancing machine, and I was able to get the big tire shop to replace the tire.

I chatted several times with this business man in the months that followed as I discovered that he was interested in and capable of addressing other car problems besides wheels. I quickly found that he could work on most any auto problem. He particularly liked the mysterious problems that often come up with today's computer controlled engines. He seemed fairly knowledgeable about many subjects and had no trouble understanding the semiconductor engineering field that I work in.

Curious about the background of such a person, I found that he is an architect by education. Because he did not find being an architect particularly interesting, he started a business in designing, installing and maintaining computer networks for companies. He became an expert at computer hardware and sold his very successful business after 10 or 15 years. Always interested in cars, he decided to learn more about fixing them. He offered to work for a friend in a car dealership for a year, for nothing, if he could attend some specialized training courses.

Although he doesn't advertise and doesn't even have a sign on his shop, his business is growing. His only computer is for engine diagnosis. He keeps appointments and accounts on paper. His dilemma is what to do next. Should he hire employees, lease a bigger shop or limit his business?

Didn't there used to be more of these kinds of people? In the past, I think every small town had such a generalist. Sometimes that person was the pharmacist, who dispensed medicine and advice on many subjects. Sometimes it was the town doctor, who was expected to know much more than just how to treat common illnesses. At times an MD was expected to be a veterinarian and vice versa. Sometimes the generalist was called the blacksmith. The blacksmith was a designer, a machinist, a welder and a repairman. The town blacksmith was probably the first auto mechanic. A couple of bicycle mechanics designed and built the first airplane. The school teacher or the newspaper editor was sometimes designated to be an expert on everything.

What happened to the generalist? Our society is two-faced on generalists. On the one hand, our educational system tries to produce specialists. Companies often will not consider hiring someone with a broad general education or broad experience. They look for someone who knows how to do one thing well. Engineering schools produce electrical engineers that are so specialized they often know nothing whatsoever about the electrical wiring in a house or the electrical components in a car. A mechanical engineer may be an expert on gas turbines and know next to nothing about a clock mechanism or the transmission in a car. On the other hand, our society seems to quietly seek out these generalists. People seem to find out whom to ask about how something works or how to solve a problem.

Maybe businesses and schools will learn to re-value the generalist. In the meantime, I'm searching for a position as a tribal elder, the ultimate generalist.

The Mysteries and Misunderstandings of Variability

Most of us have developed some distrust of statistical data, not because there is anything inherently bad about the mathematical treatment of data but because sometimes people try to use statistical techniques without understanding the built-in limitations. It is, in fact, easy to pick up one of the statistical tools and make big mistakes with it. The same is true of most tools, including the mechanical ones. I use both chain saws and nail guns from time to time. Both are quite capable of inflicting terrible injury, but only if you are unaware (or ignore) the proper way to use them. I have discovered that careful use of these tools is much easier on my body than either hammering or sawing by hand.

The misunderstanding of variability may be near the top of the list of common mistakes with data tools – sort of the "chain saw of statistics." Whenever you manufacture something and then measure it, there will be some variation. Even if the manufacturing process is excellent, small variations in the product will exist. If you look at some characteristic in a group of people, there will be lots of variation. Often we try to draw conclusions from these variations. That's good, as long as we proceed with caution and knowledge of the tools.

There are several wonderful historical examples of how you can go badly wrong by not understanding variability. The Trial of Pyx in the UK began in the 12th century and continues today as a largely ceremonial procedure for measuring newly minted coins. It began because it would be easy for the mint to short-change the king by making gold coins just a little light. The gold guinea weighed 128 grains. The people measuring chose to measure a group of coins, say 100, and they allowed 0.28

grains of variation in an individual coin. This was fine so far. Then they decided that if 0.28 grains of variation was OK for one coin, 28 grains was OK for the average of a group of 100 coins (100 times 0.28). They assumed that variability increased as the sample size increased.

In the 1900's, there was a movement toward smaller schools. To make the point with data, average student test scores in individual schools were graphed against enrollment numbers in those schools. A cursory look at the graphs clearly showed that the highest average scores were mostly in the smallest schools. The proponents made their case. Unfortunately, a closer look at the graph revealed the real story. The smallest schools *also* tended to have many of the *lowest* average scores.

The UK coin examiners and the small school proponents didn't understand the fundamental characteristic of variability. Variability decreases as the sample size increases. The average variability for a group of 100 coins should have been much less than 28 grains. A smaller school will always have greater variability in test scores than a larger school. The scores of a few sharp students or a few below average students will swing the average scores of a small group of students and have only a small influence on the average data of a large school. A more careful analysis of the same data actually showed a small advantage of large schools over small schools, perhaps because larger schools may have more resources than smaller schools.

Examples of this misunderstanding are plentiful. Every so often, someone will publish a list of the 10 safest and the 10 most dangerous cities. Without proper statistical adjustment, the list will normally have all of the safest and most dangerous cities among the smallest cities. The variability is simply higher in these cities than in the largest ones.

Luckily, some things are not much affected by variability. This keyboard gives me an X every time I type an X. XXXZXXXXX.

From the Very Large to the Very Small

If you think your mortgage is represented by a very large number, and your savings is represented by a very small number, it's probably time to put those numbers in perspective.

What's the largest thing you can imagine? OK, without a lot of thought, let's say the universe is the biggest thing we can imagine. We can even limit it to what is called the observable universe to put a few conditions on it. If that is the biggest thing we can imagine, then the biggest number we can imagine might be the number of atoms in the observable universe.

So you say, "How could we possibly have any idea how much stuff (matter) there is in the whole universe?" Without going into detail, there are actually several ways. The simplest way is to estimate the number of galaxies (like our Milky Way Galaxy) that are visible based on Hubble Telescope photos, guess at the typical number of stars in a galaxy based on our own and nearby galaxies, use the average mass of a star and the number of atoms per unit of mass, and crank out a number on a calculator. That number is roughly 1 followed by 80 zeroes.

Instead of actually writing out all those zeroes or even writing "1 followed by 80 zeroes," we use a standard shortcut called scientific notation. The number of atoms in the observable universe is 1×10^{80}. In this notation, 10 is multiplied by itself 80 times, which is the same as adding 80 zeroes to 1. This number is based on 5×10^{22} stars organized into around 125 billion (1.25×10^{11}) galaxies.

The number of cells in the human body is about 1×10^{14}. How does that compare to the number of stars in the universe? 5×10^{22} divided by 1×10^{4} is 5×10^{11} – or 500 billion times more stars in the universe than cells in body.

How about large distances? The diameter of the observable universe is about 93 billion light years (depending on what source you read). A light year is the distance light travels in one year, so a light year is 5.9×10^{12} miles. That makes the universe about 5.5×10^{22} miles or 4.65×10^{26} meters across.

How about very small things? The news is full of stories about nanotechnology, the technology that deals with things less than 100 nanometers in size. A nanometer is 1×10^{-9}, or 1 billionth of a meter. The -9 in the scientific notation simply means a decimal point is positioned ahead of the 9 zeroes. To indicate how small a nanometer is, the relationship of a nanometer to a meter is the same as a marble to the earth. The smallest bacteria are several hundred nanometers in size.

Perhaps the smallest thing we can imagine is an atom. Atoms are less than 200 picometers – 2×10^{-10} meters. Is there something smaller? Atoms are made up of electrons, protons and neutrons, so those are definitely smaller. Protons and neutrons appear to be made up of particles called quarks, which are smaller yet. Protons and neutrons seem to contain three quarks each.

Stating a size for a subatomic particle is a problem. Each particle has a measurable mass (which is related to weight), but all of these particles are so small that they behave both as particles and as waves. This makes measuring, or even defining, the diameter of these particles arbitrary. In other words, it depends on what experiment you are doing with these particles as to what diameter they will display.

In the scheme of things, humans and their daily numbers are not very big and not terribly small either.

Topics in Miscellany

Why Do We Have 12 Months and 31, 30, 29 or 28 Days in Each?

In pre-historic times, folks all over the world really needed to know what time of year it was in order to know when to plant things, when to store food for the winter, when the monsoon was coming, etc. They quickly developed knowledge of astronomical events that were easily tracked and could provide the information they needed. With a combination of the position of the rising or setting Sun on the horizon and the phases of the moon, they could roughly know which day of the year it was – long before there was a calendar. They knew that there were roughly 12 lunar cycles (new moon to new moon) in a year and 350 or so days in a year. This served well for eons of early people.

Eventually, advanced civilizations wanted to keep records, and the calendar was born. In our part of the world, the Romans were the biggest influence on the calendar we use today. Not a lot is known about the early Roman calendar, but it was probably based on a Greek lunar calendar. At first, it apparently had only 10 named months, and what we call January and February were lumped into a vague "winter period." In about 713 BC, Numa Pompilius is said to have made all of the months 29 days and added January and February to make 12 months.

The Roman calendar might have been OK except that politics prevailed. The Roman officials who kept the calendar were subject to bribes to extend or shorten months or years to the benefit of politicians. Years varied in sizes randomly. The calendar was a mess.

Finally, Julius Caesar stepped up in 45 BC. He decreed a 365 day year divided into 12 months of unequal days (because 12 doesn't divide into 365 evenly) with an extra day every four years (leap year). The Julian calendar worked well enough until the 1500's AD. Over the long term, the Julian calendar had days slightly too long because there were too many leap years. The vernal equinox (the day the Sun is directly above the equator) and, consequently, the date for Easter were slowly moving forward on the calendar. This prompted the Catholic Church to step in. The regime of Pope Gregory XIII modified the Julian calendar to try to fix the date creep.

The fix was controversial and disruptive. Ten days were dropped from the calendar to bring the vernal equinox back to its proper astronomical date. A more complicated leap-year formula was adopted. Every year that is exactly divisible by four is a leap year, except for years that are exactly divisible by 100; the centurial years that are exactly divisible by 400 are still leap years. For example, the year 1900 is not a leap year; the year 2000 is a leap year. This Gregorian calendar is basically what is used all over the world.

The next time you check your calendar, think about the centuries of churning that went on to develop it. Or maybe not.

'Twas the Night Before a High-tech Christmas...

...And all through the house,
the technology was whirring, and beeping, and glowing –,
even the mouse.
The stockings were hung on the solar heat exchanger with care,
in the hope that St. Nick would soon be there.
The children were nestled all snug in their beds,
while visions of X-Boxes danced in their heads.
And mamma in her kerchief, and I in my cap,
had just set the computers to standby for a long winter's nap,
when the outside motion detector let out such a clatter,
I sprang to the multifunction display to see what was the matter.
Away to the window I flew with an LED flash,
tore open the shutters and threw up the sash.
The Forward Looking Infrared imager on the breast of the new-fallen snow
gave the luster of midday to objects below,
when, what to my wondering eyes should appear,
but a four-hundred horse hybrid SUV pulling a sleigh with eight reindeer.
With a little old driver, so lively and quick,
I knew in a moment it must be St. Nick.
With just a glance at the GPS, he knew no shame,
He whistled and shouted and called them by name,
Now Dasher, now Dancer, now Prancer and Vixen!
Now Comet, now Cupid, now Donner and Blitzen!
To the top of the porch! To the top of the wall!
Now dash away! Dash away. Dash away all!
So up to the house-top the eight they flew,
With a sleigh full of high-tech toys, and St Nicholas too.

The solar house had no chimney around,
so poor St. Nick, an alternate route into the house he found.
A bundle of technology he had flung on his back,
and he looked like a peddler just opening his pack,
and I laughed when I saw him, in spite of myself.
A wink of his eye, a long look at his PDA, and a twist of his head
soon gave me to know I had nothing to dread.
He spoke not a word, but went straight to his work,
and filled all the stockings, then turned with a jerk.
With his finger on the teleportation button and a twitch of his nose,
up to his sleigh in a flash he rose.
To his team he gave a whistle,
and away they all flew like the down of a thistle.
But I heard him exclaim, ere he drove out of sight,
"Happy Christmas to all, and to all a good-night."

The Passing of a Very Influential Wizard

I just read that Don Herbert, "Mr. Wizard" from 1950's and 60's TV, died recently. I eagerly watched the Mr. Wizard show for a while as a kid and still remember some of the science demonstrations he performed each week for guest kids. I tried some of the experiments myself, but I was usually limited by some material I couldn't find for a good result.

This started me thinking about the chain of influence that one person can have. I didn't think much about a career in science at the time, but the seeds were probably planted. I certainly wanted to be Mr. Wizard. Don Herbert was probably at least the beginning of my passion to try to find out how things work.

Quite a few of my subsequent forays into science experimentation would not meet today's safety standards or even Mr. Wizard's standards. My gunpowder experiments were good examples, but I was also able to make fireballs with a candle and all purpose flour. The manufacturers probably didn't condone the extension of "all purpose" flour as far as I took it.

I particularly enjoyed my attempt to understand internal combustion engines. This culminated in my removing a spark plug from the lawn mower, installing it in the bottom of an empty paint can, stringing a long extension cord from the ignition of the mower to the can, and adding a teaspoon of gasoline. When I cranked the mower, the paint can lid was launched to an impressive altitude. Since I did not know exactly how the experiment was going to turn out, safety was enough of a concern that I at least used a long extension cord.

Several chemistry set experiments led to my expulsion from our house to the garage. Soon after this, I went to high school, followed by college and graduate school. The chemistry sets got steadily bigger and better. Luckily, my interests quickly turned from impressive fireballs back to understanding how useful things work.

Many years later, after I had actually studied science under more controlled conditions, the influence of Mr. Wizard returned unexpectedly. My daughter was in kindergarten in the afternoon class. The teacher made the mistake (or more likely a carefully considered strategic move) of explaining to us that the morning class was more advanced/mature and would have a science teacher come in once a week for a science demonstration. Unable to let that go by, I immediately volunteered to present a science demo to the afternoon class once a week. I enlisted the help of another parent to alternate weeks, and we embarked on an unannounced contest to be the best Mr. Wizard each week.

We pulled out all the stops. I started with a laser that I borrowed from the university (long before laser pointers were commonplace). I later heard from the teacher that one of her students was chastised by his parents for telling a falsehood when he claimed to have used a laser at school. I think the teacher was also in trouble with the morning class parents for having boring science experiments compared to the afternoon.

Our demos progressed through culturing bacteria on agar plates, dismantling a telephone, titrating a base with an acid, and then....

My final demonstration was launching a medium-sized, solid-fuel rocket from the school yard and having it descend slowly back to the launch site with a parachute. The whole school (except for the morning kindergarten class) turned out for that one.

I checked with my daughter a few years ago about her memories of that year. She only remembered the rocket. I have no information about the other students in the class, but she became a research scientist. We will never know for sure how much Mr. Wizard's chain of influence had to do with it. Anyway, I waited a little too long to thank him.

Topics in Technology

Is the Space Elevator Going Up?

Instead of powerful, expensive rockets like the space shuttle lifting payloads into orbit, how about just stringing a cable from the surface up to space? Then you just attach some kind of elevator to the cable, push the "up" button, and your payload goes into orbit.

A space elevator? Sound like science fiction to you? Actually, it *is* in some classic science fiction. As wild as it sounds, the physics of such a cable is really OK. Theoretically, you could attach one end of a cable to some sort of platform at the equator and the other end to some massive object in geosynchronous orbit. A geosynchronous orbit is one where the orbital object moves at the same speed that the Earth turns so that it appears stationary in the sky.

The fictional part of the space elevator has been the cable. No steel cable comes anywhere close to meeting the extreme strength requirements of a cable more than 60,000 miles long. Yes, sixty thousand miles. A geosynchronous orbit is around 20,000 miles, and the cable will bow significantly in the atmosphere.

Such a cable was science fiction. Now, it may not be. Along came a fascinating material called a carbon nanotube. These tiny tube-like structures can be found in specially made carbon soot. They can be formed into fibers with phenomenal strength. No cables have been built yet, but the strength is predicted to be in the range necessary for a cable to space.

Many other problems stand in the way. Communications satellites are put into geosynchronous orbit all the time, but a much larger mass would be required. Then there is the problem of dropping a cable from that orbit into the right spot. What happens if the cable ever breaks? Wind and lightning could be a problem. Other satellites would eventually hit the cable if they aren't nudged from time to time. The elevator itself needs to be a little faster than the express elevator in a skyscraper. To cover 60,000 miles in a reasonable time, the elevator probably needs to move somewhere around 10,000 miles per hour. That will require a major energy source to lift tons of payload.

These are not trivial problems, but they all have one thing in common. They are engineering problems. Americans have a history of solving major engineering problems.

When the first US manned space activity, Project Mercury, started, the engineers were just beginning to figure out how to launch a rocket into orbit without it blowing up. By 1963 they had successfully launched and retrieved a man six times. They dreamed of sending men to the moon, but only a few of them would have predicted that six years later they would be successful.

How soon can you ride the elevator? Well, NASA has already spent some money on exploring the idea. Big bucks will be required, however, to develop and install this elevator. The attractive feature is that operating costs would be much lower than rockets after the initial investment.

Whatever happens, it will be more fun to watch than most of the other things going on in the world.

How Fast Can Things Go?

At the breakfast table this morning, Grandad (as we call my 94 year old father) asked, "How fast can they go?"

He was alternately reading the newspaper and watching the hummingbirds at the feeder outside, so I asked for clarification, "Are you talking about the Olympics or the hummingbirds?"

"No, I mean what is the fastest speed that people have gone?" he asked. "In fact, aren't we going pretty fast right now as the Earth turns?"

I had no idea what the answer to the first question was, so I mumbled something about astronauts in orbit around the Earth. For the second question, I did a little mental arithmetic (after I sneaked a peek at the Internet to double check the circumference of the Earth.) The circumference is about 25 thousand miles at the equator, and we turn once every 24 hours, so that's a little over a thousand miles an hour. We would be somewhat slower at the latitude of North Carolina. It's a good thing that our atmosphere moves with us.

The first question took a little more research. Turns out that the fastest manned vehicle was the Apollo 10 capsule as it returned to Earth. This dress rehearsal mission for the moon landing by Apollo 11 reached 24,790 mph with astronauts Stafford, Young and Cernan. The fastest manned, winged vehicle is the space shuttle at 17,000 mph on re-entry. By comparison, the fastest jet-powered airplane is officially the Lockheed SR-71A (Blackbird) at 2,193 mph. Another well-known, retired supersonic plane is the Concord, which could do up to about 1,300 mph on the New York to Paris run.

When you leave out the people, the fastest man-made object appears to be the Helios 2 satellite at 150,000 mph.

This object was in an elliptical orbit around the Sun and reached this speed at its closest approach to the Sun.

The other question that we could ask is: "How fast might we go in the future?" That is actually a little easier to answer because of the speed limit. Not the speed limit enforced by the state police but the cosmic speed limit, the speed of light. Einstein taught us about a curious effect that will happen when we embark on our first interstellar flight. As the engine accelerates our craft faster and faster, and we get closer to 186,000 miles per second, we begin to get more and more massive. As our mass increases, it takes more and more energy to accelerate. The speed of light effectively becomes a speed limit that we can never achieve. (This, of course, ignores warp drives, wormholes and the other theoretical possibilities common in science fiction.)

Not terribly impressed with all of the speeds above, Grandad asked, "Well, how about the hummingbirds?"

Their speed might be the most impressive of all. The ruby-throated hummer has been clocked at 60 mph with their wings beating at 55-75 beats per second. Even more striking is that this tiny creature makes a yearly 2,000-mile round trip to a warmer climate. Not bad for a nectar-fueled bird that weighs about two paper clips.

What's All This Nanotechnology Anyway?

Underneath all the hype, a few dire predictions and some science fiction, something important is happening in the arena of the very small.

How small?

A few nanometers – or, billionths of a meter. In order to see a nanometer, you would have to magnify it by a million just to make it a millimeter, the smallest mark on your centimeter ruler. The so-called nanotechnology deals with objects much smaller than bacteria, approaching the size of a molecule or a few atoms.

So, what's the big deal? Haven't molecules and atoms been around since the beginning of the universe?

Yes, indeed, they have. Remember when I said hype? Biological organisms, including us folks, have systems inside us that function on the nanoscale all the time.

Mankind has been doing chemistry since the first fire was tended, but we are talking about more than just chemical reactions.

Although the nanotechnology term is rapidly outliving its usefulness, the Los Alamos National Laboratory tried defining nanotechnology as "the creation of functional materials, devices and systems through control of matter on the nanometer length scale and the exploitation of real properties and phenomena developed at that scale." That mouthful makes nanotechnology very broad and includes some stuff that we have already been doing. What's new is that we are learning a lot more about how the size, shape and microscopic structural detail of objects dramatically change their properties and uses. In particular, when existing materials are divided into exceptionally small particles, their properties and uses can change dramatically.

Take carbon for example. It's the stuff that makes soot, pencil lead and tires black. Recently several groups of scientists have found some very curious structures in soot. If soot is created in a specific way, it can contain carbon nanotubes. These tiny, tube-like carbon structures can be used to make fibers and other materials of incredible strength and wires with a special capability to conduct electricity. Applications include everything from bullet-proof vests and other high performance materials to electrodes in fuel cells. Fuel cells could eventually replace batteries in such portable equipment as laptop computers.

The world of the very small is certainly not brand new. Integrated circuits, the chips inside computers, cell phones and most everything electronic have had structures measured in nanometers for some years now. The hard disk drives in computers, digital video recorders and portable music players use nanostructures to store information. Integrated circuits seem poised to continue to perform more functions faster as they move deep into the nanoscale regime.

What about the dire predictions?

As with any new material, the environmental, health and safety characteristics need to be studied as part of the development research. These characteristics need to be particularly well understood as the materials move out of the laboratory and into volume production.

Separating the hype from reality won't be easy, but nanotechnology will continue to be in the headlines for years to come.

Deciphering Digital Cameras and Digital Photography

Perhaps the first question you will ask is if you need a digital camera. There are two parts to the answer. First, if you don't do computers at all, you will lose most of the advantages of a digital camera. You probably need to hang on to film for a while. However, the second part of the answer may be bad news for you. I predict that in the next five years, film will become a specialty item as well as expensive and inconvenient to use. Digital will completely take over the personal photography market. Because of this, you might want to consider making the transition pretty soon.

Just as computers have gradually taken on a major part of our business and personal communications, the digital still camera has rapidly replaced film cameras on both the professional and personal levels. What are digital cameras, and how do they work?

Most everything about a digital camera is the same as a film camera. They consist of a box-shaped body, a lens that focuses the image and controls the total amount of light that enters the camera with a shutter mechanism (variable time of opening), and an aperture mechanism (variable opening size). The body also contains a viewfinder for selecting the area you wish to photograph, a flash for taking pictures in reduced light, and some controls (sometimes just an on/off switch).

The main difference is that the film in a film camera is replaced with an electronic imager in a digital camera. The imager is a microelectronic device (chip) that is manufactured with rows and columns (an array) of tiny, square, light sensitive areas called pixels. The pixels are divided into three types by placing microscopic red, green and blue filters on top of them. This effectively creates red, green and blue sensitive pixels. Each pixel stores information related to the intensity of the light reaching it. The camera has a built-in computer that looks at the intensity information stored in each pixel. The computer looks at the information stored in clusters of three pixels: red, green, and blue. If all three pixels have equal intensity, the computer interprets this as white light in the image. If all three have very low intensity, the computer says this represents black. If only the red pixel has intensity, the image is red in that area, and so on.

In a film camera, you advance the roll of film for the next picture. In a digital camera, the electronic image from the imager is transferred to an electronic memory in preparation for the next picture. The image is really just an array of numbers that contain the color and intensity information. This information can be stored in the camera memory, or it can easily be transferred to a computer.

With this amount of information, the options available on digital cameras are easy enough to understand. The size of the pixel array in megapixels (millions of pixels) in the electronic imager is usually the primary option listed on a digital camera. These range from two to more than 12 megapixels. A pixel array of 1,600 by 2,000 pixels would be 3.2 megapixels, a common size for a point-and-shoot camera.

How many megapixels you need depends on what you want to do with your pictures. If you just want to send your pictures by email or get prints smaller than 8X10, a camera with two to four megapixels will be fine. Only if you expect to make prints 8X10 and larger or enlarge small areas of your photographs do you need a more expensive camera with a higher pixel count. The higher pixel count cameras also tend to come with more professional features that you may not need.

The professional photographer and the photojournalist were generally last in the transition from film to digital, well behind the consumer. There were several reasons for the reluctance of the pros. The simplest reason is that most pros already had major investments in film equipment, in many cases including darkroom equipment. In addition, the pros correctly pointed out in the earlier days that the resolution of digital photos could not match that of fine-grain film. In just the last couple of years, however, professional digital cameras of 12 to 22 megapixels have been available that rival and, in some situations, exceed the resolution possible with film cameras. Almost all of the other criticisms of professional digital cameras, such as ruggedness, reliability and the need for computer equipment, have been addressed by manufacturers.

Cost remains an issue for top of the line professional digital cameras, but the development of the so-called "prosumer" cameras has had a major impact on cost. The pros typically use single-lens reflex (SLR) cameras. This type of camera has interchangeable lenses and is designed so that the scene in the viewfinder is the actual image that will be captured by the camera. This is done with a moveable mirror that diverts the image from the lens to the viewfinder until the shutter button is pressed. As the shutter is triggered, the mirror snaps out of the way, and the same image goes straight to the film. Professional digital cameras are also of the SLR type. Digital SLR's (DSLR's) have been priced in the range of $3,000 - $10,000. In the last couple of years, the two major manufacturers of DSLR's, Nikon and Canon, have offered prosumer DSLR's under $1,000. These cameras offer most of the features needed by pros with a price within reach of serious amateurs.

Many professional photographers are now taking advantage of the features of DSLR's. One of these features is the wide range of effective film speeds available on DLSR's. With film, it was necessary to choose a roll with a particular ISO number (a measure of the sensitivity of the film to light) – around ISO 100 for film to be used in good lighting or ISO 1000 and higher for film to be used in very poor lighting. The photographer often carried film with a range of ISO numbers or even several cameras loaded with different films. With DSLR's, ISO can be dialed in from 100 to 1600 and higher. The same camera can shoot a scene on a brightly lit beach or a candlelit scene in a dark room, with only a simple setting change.

The electronic nature of the photos means that a photographer can send pictures in a few minutes from anywhere in the world to a publisher who has an Internet connection.

Today, such a connection is generally easier to find than film processing capability. The photographer can also store thousands of pictures on a handful of small memory cards that would barely fill a pocket.

Finally, the large investments that pros often have in lenses is not an issue. Both Nikon and Canon and several other makers of DSLR's allow the use of exactly the same lenses that were used on their film cameras.

Digital photography represents the fastest technology transition I have ever seen. This revolution was led by the consumer, who was ready for something smaller, cheaper and easier to take pictures with. Now, the digital revolution has spread almost everywhere. The last x-rays I had taken did not use film. My last driver's license photo was digital. My driver's license was ready in a few minutes, but I still need to talk to those folks about enhancement of digital photos.

At Home with Electricity

Have you ever noticed that in the movies – and once in a while in real life – a would-be murderer attempts to do in someone by tossing an electrical appliance into the bathtub with him or her? In the movies, this always works. The lights dim, and the victim is history. In reality, this is not a sure way to murder someone (other than the real possibility of scaring them to death). The science and technology related to this are interesting, and an understanding of some of the characteristics of electricity is an important element of safety in your home.

What is it about electricity that can make some areas of your home more dangerous than others? Why are we often warned that the kitchen, bathroom and outdoors are especially dangerous for using electrical appliances? If you ask people those two questions, they will often say water is the problem. Electricity and water are considered a dangerous combination.

It's true that water and electricity can be a dangerous combination, but the real reason is not always obvious. A fundamental characteristic of electricity in your home is that it will flow into anything that is electrically connected to the Earth. In the normal 115-volt house system, electricity will flow from one of the wires (the so-called hot wire) to the other wire (the so-called neutral wire) because the neutral wire is connected to Earth ground somewhere in the system. If there is an electrical appliance connected between the hot wire and the neutral wire, the appliance will operate. It may be a light bulb, an electric heater, or any of a multitude of useful appliances. If the hot wire touches the neutral wire directly, there will be a shower of sparks, and the wires will get hot. The electrical breaker or fuse is then designed to shut off the electricity until this short circuit is corrected.

The danger to us comes if we are electrically connected between the hot wire and the neutral wire or if we are electrically connected to the Earth ground and to the hot wire. When this happens the electricity will flow through us on its way to ground. It is this circumstance that can cause serious injury or death. We are warned about the bathroom and kitchen for two reasons: Water (wet hands) can make a better electrical connection to you than dry hands, and water pipes and drains in older houses are metal and connected to the Earth. Accidental contact with the hot wire in an electrical appliance by your wet hand while you are also in contact with the grounded faucet or drain will allow electricity to flow through you to get to the Earth ground. This could electrocute you. If you use an electrical device outdoors, your feet and hands could be wet and contact with both the earth and the electrical device. If the electrical device is faulty, you could be injured.

What are some of the things that give us some protection as we use electrical devices in our home, and why isn't the movie murder plot a sure thing? The answer is a mixture of technology, appliance and electrical system design, new materials for plumbing and common sense. Newer houses have three things that improve your safety around both water and electricity: 1) Three-wire electrical systems are standard. The third wire is another ground wire that is often connected to the case of the appliance if it is metal. If something happens inside an appliance, such as dropping it or flooding it with water, the hot wire will tend to connect with the case and the third wire ground and blow a fuse or breaker rather than hurting you. 2) Ground Fault Interrupters, GFI's are often connected to electrical outlets in kitchen, bathroom and outdoor outlets. These electronic devices are designed to sense very quickly (much faster than fuses or breakers) when there is a short circuit to ground and disconnect the electricity. 3) Plastic pipes for water supply and drains are becoming common because of their lower cost, durability and ease of installation. An added advantage is that they are not electrically grounded and make your kitchen and bathroom a little safer.

None of the technology above is guaranteed to prevent electrical shock, but if it is combined with common sense, you can be pretty safe around electricity.

Information Storage: The Pinnacle and the Pit

The digital age includes the capability to store extremely diverse data, everything from pictures of prehistoric cave paintings in France to music from the latest heavy metal band. (Actually, maybe those aren't so diverse, but you get the idea!) Not only can we store it, but if the Library of Congress was completely digitized, it would fit in a desk, including equipment to access it, instead of the buildings it is in now. With the proper support structure, much of which already exists as the Internet, all of that information could be available to anyone in the world at any time. So, what's the problem?

There are some big problems. First, somebody has to convert all the old stuff into digital format with cameras and scanners. This slow, costly process is actually proceeding, but there are still issues of ownership and fees that limit our access. A much bigger problem is establishing and maintaining the media and format for archival storage.

For most of recorded history after the clay tablets, the media for information transfer and storage has been some kind of paper, and the format has been one of a handful of written or printed languages. This continues today with newspapers and books. In some ways, this works well. When paper is carefully handled and stored, it can last hundreds of years. A scholar can usually be found who can read the ancient languages.

Unfortunately, with paper media, the oldest documents must be handled only rarely and only by specialists. Many documents have been lost to destructive natural and man-made forces. What we wouldn't give for a backup copy of the scrolls lost in 400 AD at the library at Alexandria, if the stories are true.

Digital storage has the potential to be resilient because the information is stored in tiny areas that are either off or on with respect to light, a magnetic field or an electric charge. This is the all-or-nothing characteristic of digital data. The storage media can be laser-written plastic discs (CD's and DVD's), magnetic media (hard drives, tapes, or floppies of various sorts), or electronic memories. Although any of these media can deteriorate with time, the digital information can be easily and quickly copied to other media. This is true unless you wait a little too long.

Therein lies the problem with electronic storage. If you have information that was written in the 1970's to an eight-inch floppy disk using the Word Star word processing software, you are out of luck today. It would difficult, maybe impossible, to find a working eight-inch floppy drive, a computer using the right bus and operating system to access the drive, and a working copy of Word Star. The data on the eight-inch floppy might still be safe, but you have no way to read it.

Today, you may be taking digital pictures. To save them more permanently or share them, you may be copying them to CD's or DVD's. This is good. The trouble comes when the next generation digs out your old CD from a dusty trunk in the attic and finds that there isn't a slot for it on their computer.

At the moment, the best defense from this dilemma is diligence in keeping your important data up to date. Data stored at the pinnacle of DVD technology will fall into the pit of media/format aging if it languishes untouched for even a few years. Libraries and research institutions worldwide suffer the same dilemma and must use the same diligence.

There may be some hope for the future. Until somebody wins the archive format battle, start training the next generation – maybe at Thanksgiving – to keep copies of the family pictures and documents in the latest media and format.

Has Anyone Seen My Laser Light Saber?

Without much fanfare, we suddenly seem to be surrounded by lasers. If you think your life doesn't include any lasers, you might have missed them. If your home or your car has a CD music player, if you have a computer with a CD drive or an inkless printer, or if you have a DVD connected to your TV, you've got lasers – all of these devices have small lasers in them. The supermarket or the discount store where you shop probably uses bar codes at the checkout. The bar code readers have lasers in them. If you have had any eye surgery or other very delicate surgery lately, a laser was probably used. Carpenters use laser levels to get molding and shelves aligned quickly in a room. They may use a miter saw with a laser-cutting guide. Surveyors use laser transits. A speaker often uses a small laser pointer to generate a small red or green dot on a projection screen. The military uses lasers for aiming everything from handheld, small arms to laser-guided bombs. They are also beginning to mark restricted airspace with lasers. Intricate metal cutting can be done with lasers. The list goes on and on.

What is a laser? Why are they used in so many devices? How do they work?

A flashlight bulb generates light when battery current flows through a wire filament. As the wire heats up, electrons in the atoms that make up the wire are moved to higher energy levels. You can think of this as electrons moving to higher (or excited) orbits around the nucleus at the center of an atom. As the excited electrons return to lower energy orbits, they give off a particle of light called a photon. In a flashlight bulb, the photons have many different energies, so the multicolored light appears white. The light is only slightly collimated – turned into parallel rays – by the flashlight reflector and lens, and the light waves are not coherent. Not coherent means, in the wave picture of light, the peaks and valleys of the light waves are not lined up. As a consequence of all of this, the beam of a small flashlight will make a broad spot a few yards away and has little application beyond illumination.

A laser generates light in a unique way. A helium-neon gas laser is an example. A tube of helium and neon gases at low pressure is excited (pumped) by a flash lamp (similar to a photographic flash). If you stopped at this point, the tube would just glow briefly, like a neon sign, as the excited gas electrons returned to lower energy orbits and gave off photons.

However, use the right mixture of gases, a flash lamp, and a mirror at each end of the tube, and a new phenomenon can occur. Excited electrons build up in orbits that allow the electrons to accumulate. A few of the electrons return to lower orbits and emit photons. These photons of light are reflected back and forth by the mirrors and are able to stimulate the sudden return of the other excited electrons to lower energy and the emission of photons. This stimulated emission triggers a burst of light, all the same color, collimated and coherent. Incidentally, that's where the acronym LASER comes from: Light Amplification by Stimulated Emission of Radiation.

If one of the two mirrors at the end of the tube is only half silvered, some of the light is emitted in a tight beam from the laser. Even from a low-power laser, the beam will produce a small spot visible several hundred yards away.

Low-power lasers have the precision to reproduce the music on your Beethoven CD or quickly read the bar code on your grocery item. These lasers are mostly solid state, laser diodes. These cousins of transistors are cheap, rugged and easily battery powered. With higher power lasers, we can do surgery, cut steel, or bounce a beam off of a reflector on the moon. The unique characteristics of laser beams are generating more applications every day.

Still a little skeptical of new technology, even Grandad put his trusted, 100-plus-year-old miter box saw in the attic when he saw the power miter saw with the laser guide. "The little red line shows you exactly where the blade is going to cut," he says with awe.

Paints, Polymers and Politics

Having been in several mostly political meetings between lawyers lately, the expression "as exciting as watching paint dry" comes to mind. On the other hand, watching paint dry can be very exciting, particularly if you know just a little about the chemistry.

What actually happens when finishes like varnish, paint and some oil finishes dry? Drying is actually a little misleading. Only a few paints work by simple evaporation of the solvent to leave a coat of some solid material on the surface. The artist's watercolors are one example of a simple system. Watercolor paints are colored solids contained in tubes or pans. They are just suspended in water and brushed onto paper or other media. When the water evaporates, it leaves behind the colors. Very simple. No chemistry. The paint just dries. A characteristic of such a simple paint system is that watercolor can be mostly wiped off the paper after it is dry. Any of the paint that has not soaked into the paper comes off easily with water.

Something very different happens if you use linseed oil or tung oil as a finish on wood. These are plant-based drying oils. The first clue that these materials are different comes if you apply one of these oils to a rag and tightly wad up the rag. In a short time, you can tell that the rag is becoming warm. In fact, a bundle of such oily rags can create so much heat that spontaneous combustion can occur. This shows that a heat-producing chemical reaction is occurring as the oil dries. Experienced woodworking shops that commonly use these drying oils and related finishes have learned to dry their rags carefully to avoid fires.

Drying oils are chemical compounds that contain lots of carbon-to-carbon double bonds. The technical term is poly-unsaturated, just like some of the edible vegetable oils that doctors want us to substitute in our diets in place of saturated fats. (The reason that unsaturated vegetable oils are healthier is another story.) As a consequence of their chemical structure, a lot of double-bond, drying oils react chemically with oxygen in the air. This chemical reaction gives off heat. The ultimate product of the reaction of oxygen with a drying oil is a polymer. Polymers are just long chains of molecules made up of the shorter units present in the drying oil. The polymer formed explains the dry, stable finish that results on wood when it is finished with a drying oil.

Almost all paints and varnishes work in a similar way. They contain a chemically reactive compound suspended in a solvent. The solvent is either water for the acrylic latex paints or paint thinner in varnishes and oil-based paints. When you apply them to a surface, the solvents begin to evaporate and the reactive compound begins to polymerize. With latex paints, the polymer is a tough, rubber-like material; with varnish, the finish is hard and transparent; and with oil-based paints, the finish is like a varnish with a colored pigment. It's clear that a chemical change has occurred in these finishes because latex paint can no longer be removed with water after it has dried, varnish and oil-based paint won't come off with paint-thinner, and so on.

I guess I need another expression for some of the boring meetings I attend. Let's see, "as exciting as watching grass grow." On the other hand, grass is pretty interesting stuff....

The Mystery of Dust Bunnies and Tiny Bubbles

It's not necessary to ponder the distant universe to find mysteries that challenge the human intellect. Just this evening I have encountered two deep technical mysteries that are a major mental exercise to solve.

The first of these enigmas is the dust bunny. Yes, the ghostly little balls of fluff that seem to appear under furniture and in corners. Why do they only form under things and in corners? Aren't those areas more protected from dust? The solution to this mystery may be at least as important as understanding interstellar dark matter.

We'll toss off the dark matter question for some other time, but let's get back to the pesky dust bunnies. Where does dust come from? That's easy. Everywhere. It comes through the window, from all of the upwind states, from your clothes, from the carpet, from the furniture, from your skin, from the dog. Everywhere. Dust is stirred up inside your house by the furnace, by walking around, by you dusting and sweeping, by everything that moves. When the dust settles, it settles on everything everywhere, only to be stirred up again by anything that moves.

Therein lies the secret of the dust bunny. The dust tends to collect and grow into these bunnies only in areas more protected from stirring. Should a reclusive bunny ever venture from hiding, he would be trampled back into loose dust. In the language of physics, dust will accumulate in areas of low kinetic energy in your living room. Map the area of lowest kinetic energy, and that's where the dust bunny will grow.

Now for the second mystery of the near universe. Tonight I extracted, with considerable difficulty, a bottle of soda from the stubborn plastic lattice that stuff comes from the store in. In the process, the bottle of lemon-lime soda was shaken up quite a bit. When I screwed off the top, you guessed it, it foamed all over the counter top. Why? What is it about shaking a bottle of soda before opening that causes this reaction? If I had let it sit for a few minutes, it wouldn't have foamed over. What changes?

The answer lies in the nature of the liquid in the soda bottle. During manufacture, the water in the bottle has multiple components dissolved in it – sugar, lemon-lime flavoring and, finally, carbon dioxide. CO_2, the chemical formula for this common gas, is dissolved in the liquid by injecting it under pressure. CO_2 is only slightly soluble in water at normal atmospheric pressure, but if you apply a little pressure, much more of it dissolves. Let a glass of soda sit out overnight, and we say it goes flat because most of the CO_2 has come out. The CO_2 is in the soda because it changes the taste. It gives sodas the characteristic bite of all carbonated beverages. The taste is partly the result of CO_2 bubbles coming out of the liquid as it flows over your tongue.

In an unopened bottle of soda, the CO_2 remains in solution because of the pressure in the bottle. As the cap is removed, the CO_2 begins to bubble out slowly. Shake the bottle either before or after opening and the CO_2 bubbles out much more rapidly, so rapidly that the bubbles push some of the liquid out the top of the bottle. If you look closely at a closed bottle just after you have shaken it, you can see that many small bubbles are distributed in the liquid. The shaking breaks some of the CO_2 gas trapped at the top of the bottle and distributes it as bubble in the liquid. When the pressure is released by removing the cap, these distributed bubbles act as sites for other bubbles to form, and the CO_2 comes out of solution much more rapidly. Letting the bottle sit before opening allows those distributed bubbles to rise to the top. Releasing the pressure then brings about a slower release of CO_2.

Before I solve any more mysteries of the near universe, I think I'll go to bed.

How Smart Are Our Electronic Computers Really?

Every so often I stop to marvel at how far electronic computing has come in only about 50 years, but computing also needs to be put in perspective. The real benchmark for computing is our brain. Where are we with respect the gray matter that sits atop our shoulders?

Electronic computers are exceedingly good at a few specific types of tasks, but they are fairly dismal failures at things that even the least brilliant of us are experts at.

The strengths of a computer are:

1) They can make all kinds of calculations at blazing speed with no errors (as long as they are provided with the correct information). This calculation expertise means that computers can rapidly sort and manipulate anything (including words) that can be represented by numbers. Computers can deal with numbers with extreme precision, limited only by the specifications (cost) of the computer.

2) Computers have vast memory which can be quickly searched and recalled without error (see 1).

3) Although they break down and require maintenance from time to time, they don't get tired or bored, and they don't need coffee.

What is our brain good at that computers do poorly?

1) Pattern recognition, integration of senses and filtering. The average person can go to a cafeteria and recognize at a glance hundreds of different foods and how they are prepared while carrying on an animated conversation with a lunch mate in a very noisy environment. This

everyday kind of event is more complex than a computer can handle.

2) Perception (related to number one). The average person can analyze subtle voice, facial and body mechanical clues to instantly tell that a friend is happy, depressed, agitated, angry, etc.. and respond appropriately. Computers are just beginning to do straight facial recognition and voice recognition.

3) Linked cognitive and motor skills. Complex skills like sports, a craft, a trade, cooking, playing a musical instrument, and creating art are typical human abilities. Computers aren't even in the ball park yet (except to manage the scoreboard).

4) The human brain requires around 20 watts of power, supplied by the food we eat, to do everything it does. Even a laptop computer requires several times this amount of power. Complex computers may never approach this low power use.

Even though computers have been developed to do parts of human activities – 1, 2, and 3 above – they don't do it nearly as well. What is even more amazing is that our brain functions well in spite of the built-in unreliability of nerve cells in our brain. Not only are our brain cells steadily dying, but we often have defects at birth and damage from strokes and injury. Our brains automatically adjust, adapt and relearn, with all of this requiring only the power of a rather dim light bulb.

Have we failed with our computer technology? We certainly haven't created real artificial intelligence, but it's hard to say that we have failed. Today's computer technology is good at exactly the things that we humans are the worst at. That's what makes them so useful. We are using computers as I think they are intended – as tools. These tools will continue to improve. Computers will get closer to human abilities, but we don't need them to be athletes, artists or writers. We need them to take care of routine things so that we have time to be athletes, artists and writers.

I'm happy that this little machine has, in the last minute, corrected my spelling of athletes (three times), told me that I have 570 words and 2,822 characters so far, and that I have a mail message from my daughter.

There Is More Chemistry in the Kitchen Than You Probably Thought

Ever hear that rubbing your hands on stainless steel will remove the smell of onions? Ever wonder what it is that makes you cry when you cut onions? Do you hear that baking soda will help clean a pan or a drain, that baking soda or ammonia will remove odors from the air, or ever wonder what the difference is between baking powder and baking soda? The explanations of all of these things have one thing in common: chemistry.

Onions are wonderfully complex in their chemistry. The volatile compounds that give onions and garlic (and everything in the *Allium* genus) their taste and odor are amino acid sulfoxides. These compounds are not the culprits in tear production, however. When onions are crushed or cut, some enzymes are released that convert the sulfoxides to compounds called lachrymators (mainly syn-propanethial-S-oxide). A lachrymator is a substance that causes your eye to tear. Interestingly, these enzymes are not present in garlic, so the characteristic taste and odor of garlic are there – but no tears.

I don't know of any scientific study on the effect of stainless steel on onion/garlic smell on your hands, but it seems to work to some extent. The likely explanation is that the surface of stainless steel is somewhat microscopically porous and could probably chemically adsorb compounds like amino acid sulfoxides. We have even seen a stainless steel bar shaped like a bar of soap for sale in a gourmet cooking store.

Baking soda is a simple chemical compound, sodium bicarbonate, which has been known for centuries. Its uses are so numerous that we can't cover all of them here. Its usefulness is primarily due to the fact that it is chemically a weak alkali, or base. As such, it can do two things: It can react with and neutralize acids, and it can react slightly with grease to form a soap.

The odor absorbing capability of baking soda in the refrigerator, in the kitchen, as a foot powder, etc., is due to the fact that many odors are the acidic by-products of bacterial action. The chemical reaction between the acids and the baking soda can reduce the odor. Ammonia is also a base, and it is said that a dish of ammonia in the kitchen will absorb odors. It is also a well-known cleaning agent. The trouble with ammonia is that it has its own odor.

The cleaning capability of baking soda is also related to being a base. Bases can react with greases and oils to form soaps. In the process of doing this, baking soda can loosen baked on grease in a pan or a greasy clog in a drain. When baking soda reacts with acid and water, it foams as carbon dioxide gas is released. The foaming, bubbling action also helps the cleaning process.

The bubbles of carbon dioxide that are released from baking soda when it reacts with acid are the reason for the main use of baking soda in baking. The combination of baking soda with an acid in a dough recipe makes the dough rise from the carbon dioxide bubbles. Yeast does the same thing in a biological process rather than a purely chemical one.

Finally, we come to the difference between baking soda and baking powder. Baking powder is just baking soda mixed with one or more powdered acids. If it is single acting baking powder, it will bubble as soon as it is mixed with water and make your recipe rise at room temperature. Double acting baking powder adds another acid which forms at higher temperature and makes the recipe rise as it bakes.

Light Emitting Diodes Are Not Just Red Anymore

LED's, the acronym for the diodes that have been common for years as the little red indicator lights on electronic devices, have been undergoing a quiet revolution. They now come in many colors, including white, and they have become much brighter. Because of these two changes, they are showing up in many more applications beyond indicator lights. The two seemingly small changes have created a genuine lighting revolution.

What is an LED? The answer is in the name. They are a special kind of diode. Diodes are solid state semiconductor electronic devices, which have the characteristic of passing current in only one direction while blocking the current in the opposite direction. A light emitting diode is specially designed so that it emits light as it passes current. The details of the complex chemistry in the semiconductor manufacturing process determine the color of the light emitted. Red turns out to be the easiest LED to make. That's why LED's have been mostly red since their invention in the 1950's

In the past few years, some even more complex manufacturing processes have been able to produce LED's in many colors that are many times as bright as earlier LED's. Now that they are also very bright, LED's have built-in characteristics that make them desirable in some applications. Efficiency and durability are the most important characteristics. Because these devices are much like transistors and very unlike light bulbs, they produce very little heat and have lifetimes in the 10-year-plus range. The lack of heat production means that they are much more efficient – that is, they produce much more light per unit of electricity – than any light bulb. In a battery powered flashlight with LED's instead of a bulb, that translates to batteries lasting many times longer. The long lifetime of LED's compared to bulbs is a result of the inherent shock resistance of all transistor-like devices and the low operating temperature.

Where are these new LED's showing up? One of the first places that you may have already noticed them is in traffic lights. If you look closely at many traffic lights, they are made up of many pinpoint lights rather than a single bulb. Each of those pinpoints is an LED. Now, when a traffic light is out, it will never be due to a burned out bulb. You also see bright red LED's in many car and truck tail/stop lights and in various lights on emergency vehicles. A wide variety of lanterns and flashlights with LED's are available with long battery life and extreme ruggedness.

When do these LED's replace all those pesky light bulbs in your house that always burn out in the most difficult spots to reach? There must be a catch. The new LED's are expensive. Their cost has been coming down, but they still only make economic sense in applications where reliability is very important, bulb changing is difficult and expensive, or battery power is important. Stay tuned. As LED manufacturing costs continue to come down and the cost and consequences of energy use continue to go up, light bulbs may still go away.

Just think, one whole area of non-politically correct humor will go away, too, if light bulbs never need to be replaced. You know, "How many (name a group) does it take to change a light bulb?"

If You Ever Need a Bulldozer on the Head of a Pin

Microelectromechanical Systems (MEMS) are microscopic machines that do useful tasks. If you are thinking of just futuristic, laboratory stuff, think again. You have likely already used some of these devices.

Inexpensive inkjet printers use a MEMS device as part of the print head to generate and direct microscopic drops of ink toward the paper. Most late model cars are equipped with airbags. The sensor that triggers deployment of the airbag in a crash is a tiny MEMS device that detects the sudden acceleration change. If you have used a conference room computer projector, many of them have a MEMS device made up of thousands of tiny mirrors to create and project an image. Some projection TV's use a similar device. Some higher-end cars have a dynamic stabilization system to prevent skids that combines braking and power application to drive wheels based on a MEMS gyroscope. The list goes on.

MEMS are cousins of electronic, integrated circuits (IC's) that power almost all electronics from computers to cell phones. IC's are made from super-pure, single-crystal silicon that is treated with traces of certain elements to make it selectively conduct electricity. The crystal is then micro-machined to create millions of devices called transistors that switch or amplify electric currents. A MEMS device, on the other hand, focuses on the micro-machining part of the process and is more mechanical and less electronic. Instead of transistors, a MEMS process produces tiny motors, gears, valves, actuators or other moving parts. MEMS devices are usually based on silicon, but they – like IC's – can also include metals and other materials. Some MEMS devices actually combine electronic and mechanical elements.

The futuristic, laboratory stuff could be really wild. You might be able to make tiny robots that could enter your bloodstream and repair blood vessels, attack cancer, make repairs at the cellular level, and then self destruct. The possibility of MEMS devices interfacing with nerves is being explored as an approach for sight for the blind and movement for the paralyzed. You can imagine tiny factories that could construct tiny devices that can make even smaller devices. The military imagines "smart dust" that can be sprinkled around a battlefield. The self-contained "dust" robots draw power from their environment, communicate with each other, and organize themselves into groups that can observe and communicate details of enemy activities.

When MEMS become even smaller, they become NEMS or nanoscale MEMS. Nanoscale usually relates to less than 100 nanometers, which means you would have to enlarge a device 10,000 times to get it to a millimeter, the smallest mark on a centimeter ruler. NEMS may approach the sizes of atoms and molecules. At this scale, materials begin to have unusual properties. Instead of micro-machining techniques used for MEMS, it may be possible to use self-assembly properties to build NEMS. Because nanoscale elements of NEMS are greatly influenced by electrostatic forces rather than gravity, the possibility arises of using biological or biological-like membranes as templates to assemble NEMS devices. Laboratory demonstrations of this self-assembly approach are appearing in scientific literature.

The science of nanostructures is still young, but already important materials have emerged. Carbon nanotubes, tiny tubular structures that can be isolated in soot, are proving to have very useful properties. When added to other materials, they can produce fibers much stronger than steel. Carbon nanotubes also have electrical properties that may mean they can be used to make very tiny electronic devices. Of course, as with all new technology, you tend to hear more about the unknown, scary aspects of nanotechnology rather than the useful ones.

116

We Take a Shellacking and Don't Even Know It

I must admit a continuing fascination with wood finishes. The science of the finishes is interesting enough, but I am always looking for an easier, quicker way to finish a project in wood. This brings up French polishing. This old technique is almost the opposite of my desired process. It involves a laborious process of applying hundreds of layers of shellac with a small pad. It reminds me of spit-shining shoes, a process I learned well in the US Army and have avoided since. It is decidedly not quick or easy, but it involves a fascinating material called shellac.

Shellac is derived from lac, a natural polymer that has never been successfully synthesized by man. Lac is secreted by a small insect (Laccifer lacca) on certain trees in India and Thailand. (An aside: Although the term lacquer apparently derives from lac, lacquer is a completely different material that predates and does not contain shellac.) The polymer is secreted by the female insect as a hard, protective cocoon for its larvae. Small, encrusted twigs, called "sticklac," are gathered by hand and combined for purification. Trees are carefully cultivated and "inoculated" with "seed" insects to keep the process going. This lac material has been collected for around 3,000 years.

In early times, the ancient Chinese and Indians extracted the reddish dye from lac to dye silk and leather. The material was also melted to mold various objects or used as a hot-melt glue to set jewels or attach the hilt to a sword.

By the 17th century, artists in Europe and elsewhere were using a form of lac in their paints and as a protective coating on their paintings. About the same time, furniture makers began to use the same material to provide a durable finish on their creations. This form of lac received an additional purification step. The lac was placed in fabric bags shaped much like a fire hose and rotated over a charcoal fire. When the lac melted, the bag was twisted to squeeze the lac through the pores. The material on the outside, called "shell lac," was scraped off. The process removed additional impurities; gave a harder, lighter colored product; and provided the common name, shellac.

Beginning in the early 1900's, the uses for shellac multiplied almost explosively. An additional purification step – bleaching – left the shellac flakes colorless and even harder. Most of the shellac was purified in fairly modern factories. Shellac was used as a binder in the manufacture of gramophone records, shoe polish, felt sizing for men's hats, hair spray, floor wax, printing inks, adhesives, paper and foil coatings, grinding wheels, and coatings for everything from pills to candy. Up through the 1950's, home builders and painting contractors used shellac as a sealer for plaster walls and as a fast drying varnish for interior woodwork.

Interestingly, the development of modern varnishes and synthetics almost put an end to shellac in the industrial coating business by about 1960. The exception was the non-toxic market. Since shellac is completely non-toxic, it is still quietly used in many products: candy coatings, enteric coatings on pharmaceutical tablets and time release capsules, and coatings on fruit. Reportedly, insiders in the coating trade refer to shellac as "beetle juice."

In spite of a general decline in the use of shellac, there are still devotees, and there appears to be some resurgence in its use. Fine guitar makers still prefer the French polish technique with shellac for the best tone from their instruments. Specialty furniture makers still use shellac in several stages of their work. Shellac is easier to refinish because it is easily removed with alcohol and reapplied.

The best solvent for shellac is pure grain alcohol from the local liquor store. This makes shellac solutions the lowest toxicity of any finish, a feature of increasing importance with respect to air pollution, finishes for eating utensils and toys. Of course, you might not want to sip your shellac unless you are absolutely sure it is not dissolved in denatured, poisonous alcohol.

New Data Source for Illegal Drug Use and More

Cocaine use is a very serious, worldwide drug problem. One of the issues faced by law enforcement agencies is obtaining accurate information about the level of the problem in a particular area. They obtain information from fatalities, people seeking rehabilitation, arrests, drugs seized, etc., but none of this information gives an accurate, up-to-date picture of the problem in a particular area. Enforcement agencies do not have a good way to measure how effective their efforts are, particularly in the short term.

Chemistry and some innovative ways of environmental sampling may provide a better way of monitoring this particular drug problem.

In Europe, sampling the money may help to check on cocaine use. It turns out that the high quality paper used for currency is particularly good for absorbing and retaining small amounts of cocaine. The cocaine probably gets into the money because of the tradition of cocaine users to snort the cocaine powder through a rolled up bill. These individual bills can then contaminate other bills that they come into contact with as they are handled. Modern electronic instrumentation for chemical analysis can detect tiny traces of cocaine on bills.

Many European countries have currency that tends to circulate within the country. Paper money has, on average, a lifetime of about a year before it is destroyed and replaced with new currency. Over time, data on samples of a country's currency can give a picture of cocaine use in that country. The first experiments with this technique in several European countries have collected data that correlated well with other traditional data.

The money sampling approach has value in smaller countries, but it can only give a general overview of the country. Currency data in the United States would not be very helpful because paper money can be transferred long distances from bank to bank during its lifetime. What is needed is a more area-specific technique that provides quicker response.

When cocaine enters the human body and does its thing to the brain, it quickly exits the body through the kidneys in the form of a breakdown compound, a metabolite with the imposing name of benzoylecgonine. This is the stuff that periodically gets pro-athletes and others in trouble when it shows up in their urine and causes them to fail a drug test.

This chemical compound is unique in the environment and is fairly stable. So, you can guess how it might be possible to monitor a city, or a section of a city, or any collection of dwellings that are serviced by a sewer. The data can be real time, and the main uncertainty is just estimating the relationship of the measured benzoylecgonine concentration in sewer water with the number of users and level of use.

As this technique comes into use, there will be many political and legal squabbles. Already, at least one US city has refused to allow sampling of its sewer water because of fear of losing some of its federal funding. Perhaps they aren't doing as good a job of curtailing cocaine use as they have claimed.

The analysis of sewer water has surprising potential. There are literally thousands of chemical compounds that could be detected in a sewer. It might be possible to generate data pictures of cities and neighborhoods with respect to pharmaceutical use, home products used and even food preferences. I could even imagine a court order for sewer tapping instead of wire tapping. The paranoid among us may return to the outhouse and padlock it.

An Uncommon Competition for the Higgs Bosun

The winner of this race won't make the headlines of a lot of newspapers, and very few people have even heard of a Higgs bosun. Interestingly, billions of dollars have already been spent on the race to find Higgs. The US is participating in the race but is not favored to win.

When I studied science in high school in the 1960's, we learned that atoms were made up of electrons, protons and neutrons. Things got much more complicated after that. Only the electron remains a fundamental particle (can't be broken into a smaller particle), and there are now 15 particles in addition to the electron. In about 1970, these 16 fundamental particles made up a theory in atomic physics called the Standard Model of Particle Physics. Some of these particles were given whimsical names by the physicists that discovered them – including the six quarks which were labeled up, down, bottom, top, strange and charm quarks.

The Standard Model of Particle Physics is a tidy model of atomic physics except that one piece is missing. The missing entity is called either a Higgs field or a Higgs bosun, depending on whether you were talking about fields or particles. Basically, the Higgs field is the thing that allows many of the 16 fundamental particles to have the property of mass. (Mass is the property that allows materials to have weight.) In a nutshell, Higgs gives us mass, and gravity gives us weight (and ice cream increases both). Without Higgs, the Standard Model of Particle Physics doesn't completely hang together.

The Higgs field/Higgs bosun was proposed by Professor Peter Higgs of Edinburgh University. It has unofficially become part of the Standard Model, but it has remained elusive as far as definite detection. Many physicists have been trying. Groups at Fermilab in the US (near Chicago) have been using their accelerator, called the Tevatron, to look for Higgs without success yet.

Accelerators, commonly called "atom smashers," are huge, usually circular, very expensive pieces of vacuum equipment. Any particles with electric charges (electrons, protons, ions) can be steered into a circular beam, accelerated to very high energy, and made to collide with other particles. The sub-atomic particle debris from the collisions can be analyzed to find new particles as well as find processes by which particles interact or even interconvert.

The US began a project called the Super-Conducting Super Collider (SSC) in 1987, funded by the US Department of Energy and located in rural Texas. Even though a couple of billion dollars were spent, and miles of tunnel were dug, the US Congress cancelled the project in 1993. This probably removed the US' best chance of finding Higgs and certainly discouraged the worldwide physics community for a while.

Now, the best hope lies on the border of France and Switzerland with a project called the Large Hadron Collider (LHC). The organization behind LHC is called CERN and is made up of European members and global affiliates, one of which is the US. (Affiliates are passengers, not drivers.) LHC consists of a 27 km diameter (16.2 mile) accelerator ring in a circular tunnel along with the necessary super-cooled magnets, sophisticated electronic detectors and large computers.

So, why do we care about all this theoretical physics? Even if we put aside man's deep-seated need to explore and understand our surroundings, learn how the universe was born, and predict what will happen to it in the future, technology is quietly driven by big projects like the space program and collider projects. Without them, we wouldn't have today's communications, computer and electronics industries. While we may sometimes long for simpler times, most of us would miss the prosperity and the ease of access to information, communication and entertainment that has been generated.

With luck, LHC will fire up soon. If they find Higgs, it will make a lot of physicists happy. If they don't find it, physicists will have to make a new Standard Theory and a bigger collider. That will make them even happier.

Don't Be Surprised if Your Compass Points South One of These Days

Because of an action called sea-floor spreading, which could be a topic of its own, iron containing basaltic lava flows out of cracks under the ocean at a reasonably constant rate. As soon as the lava cools into rock, it locks in a record of the Earth's magnetic field at that time. When these undersea lava flows are studied with sensitive devices to measure their magnetic properties, they seem to fall into a striped pattern. Each stripe has opposite magnetic polarity.

This information clearly indicates that our Earth's magnetic field flips every so often. After a magnetic pole reversal, compasses point in the opposite direction.

The first question of interest to us is: "How often does this flip happen?" The lava record indicates that the last time was over 750,000 years ago. This was during the middle Pleistocene Era. Neanderthal man roamed the Earth along with mammoths and saber tooth tigers. An ice age dominated the period. Man probably hadn't discovered the compass yet, so he might not have noticed the change.

The lava record indicates that reversals have occurred at intervals from tens of thousands of years to millions of years. The average is around 250,000 years, so we might be due.

The next question is, "What causes a magnetic pole reversal?" That is a more difficult question. We know our magnetic field is generated in the Earth's core, but we actually know more about outer space than we know about the center of the Earth.

Space is easier to explore. We receive information from the entire electromagnetic spectrum with telescopes of all sorts. We have sent robotic probes all the way to edge of our solar system and men as far as the moon.

We only have a type of sound wave, called seismic waves, to explore the center of the earth. Probes have only gone deeply into the crust of the earth, nowhere near the core. Our information is really only based on seismic studies when powerful earthquakes send out waves that traverse the core. From those studies, we think the core is made up of iron-nickel. The outer core appears to be molten, but the center is under such high pressure that it appears to be solid.

The current theory says that convection currents and currents generated by the rotation of the Earth in the molten outer core create a giant electromagnet that roughly aligns with the rotational axis of the Earth. Computer simulations and magnetic field measurements on the Earth's surface suggest that these currents of molten metal are not uniform or completely stable. Islands of reversed magnetic polarity develop and either dissipate or begin to dominate. On a random basis, the overall polarity may reverse as one of these islands grows. The computer simulations suggest that the reversal is only sudden on the geologic time scale and may still require from hundreds to thousands of years to complete. During this time, the Earth's magnetic field may be weak, fluctuating, or it may disappear entirely.

It is expected that the magnetic poles would drift away from the rotational axis just prior to a reversal.

Our current status is that the Earth's magnetic field is weakening at a rate that would have it disappear in about 3000-4000 AD, and the north magnetic pole is drifting toward Siberia.

So, if the magnetic field disappears and then reverses around 4000 AD, are our offspring in trouble? Fortunately, the answer is probably not – at least, not from this. There could be some chaos as the Earth's protection from the solar wind disappears, but there appears to be no correlation between mass extinctions (like the disappearance of the dinosaurs) and field reversals.

Flapping to the Future with Ornithopters

Today, if you look out the window from your airplane seat, and the plane is flying by flapping its wings, you might want to pass up the next round of those little bottles that flight attendants sell, get off at the next stop and check with your family physician. Sometime in the future, however, you might be OK and just be flying on an ornithopter.

The idea of an ornithopter, a plane with movable, bird-like wings, has been around for as long as man has been watching the birds soar. Leonardo da Vinci devised several flying machines based on bird wings. Unfortunately, Leonardo never was hooked up with someone skilled in building things, so his ideas were not tested. Many other inventors have tried to copy the wing-flapping approach to flying. You have probably seen some of the grainy films of hapless pilots trying of get off the ground with their flapping contraptions that either self destructed or just didn't fly.

Currently, we are able to soar something like a bird with a variety of gliders, but all of these devices depend on a tow from some vehicle, an engine to push or pull the glider, or a big hill to jump from. Some of the glider wings can be warped to change direction or lift, but none of them provide their own lift from wing flapping.

The reason for a long history of flapless designs is simple enough. The duplication of a bird's wing is mechanically complex unless it is a fixed wing just for gliding. Man also wasn't designed with strong enough muscles, compared to his weight, to flap wings, even if he had them.

Now, the situation with wing flapping may be changing.

Many universities and other research organizations worldwide are looking at shape-change aircraft. The

University of Toronto has "flappers" ranging from one-fourth scale models to a jet powered, full-sized plane. The models fly well. The full-sized plane flew only briefly a few feet off the ground and then landed roughly.

An even more exciting approach has been developed by researchers using a composite wing. The top layers would be solar cells, intermediate layers would warp with an electric current to flap the wing, and an additional layer would be a polymer-based battery to store current. This design has attracted the attention of NASA for funding. Once developed, a plane based on this approach could be used for extended atmospheric observation on a planet like Venus or long term, un-manned scientific or military observations on Earth. Such a plane would be perfectly quiet, look very much like a bird, and stay in the air indefinitely.

If you are a model aircraft hobbyist, the ornithopter is a popular design for experimentation with everything from rubber-band power to fuel powered, radio controlled models available.

For now, if it's up in the sky and flapping its wings, it's most likely still a bird. A decade from now, who knows?

What Do Whales, Naval Personnel and New Parents Have in Common?

Picture a sailor in the sonar room of a World War II destroyer trying to locate an enemy submarine with active sonar. His equipment emits a characteristic metallic PING, and he waits for a return echo from the hull of the submarine. It's a tense moment. The sailor probably never imagined that the same technology was going to be involved in another tense moment years later in a completely different setting. Nervous parents of the sailor's grandchildren are looking anxiously at a computer screen in the doctor's office as a technician moves a handheld probe over the abdomen of the new mother. They are relieved to see an image of a small, but healthy-looking, baby boy.

In a way this technology on a war ship and in a doctor's office was a long time in coming. For millions of years, whales and various other undersea creatures have used sound-based, echo locating to stay out of danger, keep track of each other and find food. Whales can emit a range of clicks, chirps, and groans to locate objects in their dark, sea-water world as well as communicate over significant distances. It was only about the time of the First World War that man got serious about undersea sound.

Because of the threat of submarines, the British made extensive use of undersea microphones – hydrophones – to detect submarines. This listening approach is what we would call passive sonar today. The technique is still a major part of naval technology. Soon after the passive technique, active sonar was developed. This powerful technique allowed not only the detection of an underwater object, but the measurement of the distance to and velocity of the object. The approach sends out a burst of sound, a ping. By measuring the time required for the echo from an object to reach the microphone, the distance can be calculated.

Medical diagnostic ultrasound was first used in the late 1940's at, as you would expect, the Naval Medical Institute using naval sonar technology. Since that time, the medical sonogram has come a long way – all the way to the point where an ultrasound image of an unborn child is a fairly common part of prenatal health care.

Medical applications require a much higher frequency than naval sonar because the size of the objects involved and distances to them are much smaller. Ultrasound refers to sounds above the range of human hearing, about 20,000 hertz (cycles per second). Medical ultrasound is, in fact, much higher than that, in the range of one to 15 megahertz (million hertz). This is actually the radio frequency range applied to sound.

Medical ultrasound equipment is usually portable and mounted on a cart. The handheld, ceramic transducer is both a speaker and a microphone for the ultrasound. A water-based gel is used to achieve a constant coupling between the transducer and the skin. The transducer emits short bursts of sound and listens for the return. Based on the time delay of the return and the loudness (amplitude), a computer creates an image. The brightness of the image is related to how hard or soft the tissues are. The sound tends to be completely blocked by air and bone, so seeing inside intestines, lungs and brains is difficult.

A more recent enhancement to ultrasound imaging is the Doppler ultrasound. The Doppler Effect is just the change in the pitch of a sound depending on whether the source is moving toward you or away. The Doppler ultrasound can present a color image showing blood flow. This can allow a doctor to detect blood clots and other blood flow abnormalities.

Medical ultrasound is considered a safe, non-invasive tool and naval sonar has a long history. As with all technology, however, there are trade-offs. Medical ultrasound does transfer energy (mechanical and heat) to tissues and should probably only be used when really necessary. Active sonar has undesirable effects on whales and other sea life. At the very least, sonar to a whale is like our neighbor with the too loud radio. Perhaps the whales will band together one of these days and blast the headphones off a sonar operator.

Has Free Energy from the Sun Arrived?

The idea has always been simple and appealing: Put some kind of solar collector on your roof, and use the almost unlimited energy that streams from our Sun. In some areas of the world, the skies are clear almost every day; and most everywhere, except extreme northern and southern latitudes, there is enough Sun to be useful all year. (If you like odd facts, there are 3,850 zeta joules per year of solar energy available to the Earth.) Our Sun is an ordinary main sequence star that should be stable for a few billion years. Solar energy appears to be completely emission-free, the ultimate green technology. So, why aren't we all using it?

Unfortunately, the answer has historically been: It's expensive. The startup costs are high, and the design, construction and maintenance can be complicated. Just finding a qualified contractor has been difficult in some areas.

Recently, things appear to be changing in a curious way. Solar panels are increasingly appearing on the roofs of commercial buildings. Some of the big retail stores and other corporate buildings are capitalizing on their expansive, flat roofs by having solar arrays installed.

California seems to be leading the way with about 70% of the electrical utility grid connected solar arrays in the US. It turns out that the climate in California – not just the sunshine but the government subsidy climate – is very favorable to solar energy. The installation costs of electrical solar arrays have been coming down but are still high. A system will eventually pay for itself, but government subsidies shorten the payback time.

Corporations have also found that they can creatively finance solar arrays. Small utility companies have formed that bear the cost of equipment installation and then sell electricity back to the business during peak business hours at guaranteed rates and sell the rest on the general utility grid.

Solar arrays come in two general types, electricity generating photovoltaic cells and thermal transfer arrays. The thermal array is more common on residences. These panels heat water for hot water and for cold weather interior heating. A backup system using electricity or fuel usually takes care of nights and cloudy days.

Photovoltaic cells produce direct current (DC) electricity. The DC is usually converted to alternating current (AC) by an electronic device called an inverter. AC current can then be used in lighting and appliances or connected to the utility grid and sold to utility companies.

The big question is if the worldwide surge in photovoltaic arrays on corporate buildings in the last five years is a temporary fad. Will it wane when subsidies expire and the corporate interest in having a green image slows? It may not. At the current rate of solar cell cost decline of about 5% per year, solar electricity will match other sources on the grid about 2015. The issue of global warming and climate change will also continue to heat up, putting more pressure on green house gas emission reduction.

Although the total environmental impact of solar cells is favorable, it is certainly not zero. Photovoltaic cells have to be manufactured. They are manufactured in a semiconductor plant, much like transistors and other electronic components. The manufacturing plants are low emission, but they use significant amounts of energy and water. Since the bulk of energy is still generated by burning fuels, the making of photovoltaic cells still requires burning fuel and using other resources. In short, energy from the Sun is still not really free, but it will be interesting to watch.

In spite of some unresolved issues with solar power, the idea of the watt meter on the side of my house running backward and payments coming to me from the power company is an appealing idea.

Nukes Have Aging Problems Just Like the Rest of Us

Picture the following: A huge stockpile of thousands of nuclear weapons stored in various places around the world and carried by ships, submarines and aircraft. The warheads were all built in the late 1970's and 1980's with design lifetimes of 20 to 25 years. None of the weapons have been tested since the early 1990's, when underground nuclear testing was stopped in the US. Not only do all of these systems have normal aging of their complex components, but all of these weapons contain radioactive materials that accelerate the deterioration of the materials that are exposed to the radiation.

Although most US leaders have expressed an interest in eliminating all nuclear weapons eventually, they have also agreed that until the whole world agrees on scrapping nuclear weapons, the US must maintain a nuclear deterrent. If the weapons had been left to deteriorate, the US nuclear deterrent strategy would be completely ineffective by now. The likelihood that an individual weapon would actually yield a nuclear or thermonuclear explosion would be low.

From 1945 through September 1992, the US maintained an often secret but extensive nuclear weapons development and manufacturing strategy. New weapons were designed at centers like Los Alamos and manufactured at several locations, including Oak Ridge and Hanford. Older weapons were replaced. An important part of this engineering activity was testing. At first, the weapons tests were above ground in places like Bikini Atoll in the Pacific. This era ended in 1962 and testing moved to underground locations at a Nevada test site. Underground testing continued until 1992, when a moratorium on all nuclear testing halted them.

Beginning in 1992, the government's weapons experts moved into a unique mode of operation. They were chartered to maintain the US nuclear deterrent while building no new weapons and conducting no tests. This created the current approach of extensive inspection and maintenance of the weapons cache. As parts, or even larger systems, have degraded or become obsolete, they have been replaced. This is a complex and expensive process, and the question always remains just how long a nuke can be maintained in ready status.

Not only must parts be replaced, but in some cases, the nuclear fuel must also be replaced. In a thermonuclear bomb (popularly called a hydrogen bomb), the material for the fusion reaction is tritium, a radioactive isotope of hydrogen. Tritium has a half-life of 12.3 years, so it must be replaced regularly. Luckily, plutonium-239, the radioactive material used as the primary in a hydrogen bomb, has a half-life of 24,000 years. It doesn't need replacement.

A debate is also underway on the effectiveness of maintenance versus replacement. Some officials argue that the stockpile must be replaced with new weapons to be an effective deterrent. The designs and technology for building nuclear weapons have improved greatly since the 1980's. The new weapons can be simpler, more reliable, easier and safer to manufacture, and less expensive to maintain.

On the other side of the argument, geopolitical forces are against building any new weapons of mass destruction. New nuclear weapons (even if they replace old weapons) would probably violate various treaties we have signed over the years and would compromise our influence on obtaining concessions from the other nuclear powers. I tend to fall on the side of maintaining the old ones until we can scrap them all, but stay tuned.

Topics in the History of Technology

Digital Cell Phones, Torpedoes and a Movie Star

What do these three things have in common? They are connected by one of the best stories in the history of technology development.

Beginning the story in the present, many digital cell phones use a radio technology called spread spectrum. There are several ways to do spread spectrum radio. The simplest to understand is frequency hopping. Instead of your cell phone sending your digitized voice on a single radio frequency, and the cell tower receiving it on that frequency, the phone splits your conversation into short segments and transmits them on a series of random frequencies. As the transmitter hops from one frequency to the next, the receiver is synchronized to the same hops. This technology mainly does two things. It allows many cell phones to operate on the same band of frequencies with minimal interference, and it offers some extra security for your call. You can imagine that it is hard to eavesdrop on a radio signal that is constantly changing frequency.

Some years ago when researchers became interested in frequency hopping, they discovered that US patent 2,292,387 had been issued in 1942 to H. K. Markey and George Antheil for a "Secret Communications System." The patent describes frequency hopping and suggests a mechanical mechanism with punched paper tapes, like a player piano has, to implement it. Possible uses of the system included radio controlled torpedoes. The idea of a torpedo controlled by a radio signal from a plane or ship was not new, but it had never been implemented because of the danger that the signal might be jammed or replaced with another signal.

138

It turned out that H. K. Markey was the married name of the Hollywood movie star Hedy Lamarr. George Antheil was an avant-garde composer. So, how did a movie star and a musician come to invent and patent spread spectrum radio more than 30 years before it was used?

Hedy Lamarr was born in Vienna in 1914 as Hedwig Eva Maria Kiesler. In 1933, she married Fritz Mandl, one of the five leading European armament manufacturers. Mandl specialized in shells and grenades, but from the mid-1930's on, he also manufactured military aircraft. He was interested in control systems and conducted research in the field. Mandl kept his young wife by his side as he attended hundreds of dinners and meetings with arms developers, builders and buyers, where Lamarr clearly learned some things. The marriage broke up in 1937 when Lamarr escaped to London. She soon was signed by MGM, where she became a popular US movie star under her stage name.

In 1940, Lamarr became friends with George Antheil, an avant-garde composer with an interest in synchronized player pianos. Their conversations at some point turned to frequency hopping and its possible use for radio controlled torpedoes. Her interest in the subject probably resulted from the discussions she had heard between Nazi customers and her husband. Antheil thought it might be possible to implement the technology by using paper tapes. The two sought to obtain a patent and to interest the US Navy in the technology for application in torpedoes. The patent was issued, but the Navy passed up the idea. They apparently couldn't figure out how to get the piano into the torpedo.

Years later, the transistor made frequency hopping easier to implement. Then, even later, the digital cell phone and other digital wireless technologies needed the characteristics of spread spectrum radio. Now, it is used routinely. Of course, all of this happened after the Lamarr/Antheil patent had expired, so neither of them ever received a nickel from the invention.

So much for the idea that all of the important inventions come from scientists and engineers.

There Shouldn't Be Very Much Technology in Music, Right?

I decided recently that my non-technical education is lacking. It's time to work at something that does not depend on technology. Music seemed like a good choice. Something simple, like a guitar, seemed like a traditional, inexpensive, portable and easy-to-learn instrument.

I was tempted to build the instrument. Books, tools and supplies abound for the amateur luthier (stringed instrument builder). After studying luthier literature with great excitement, I decided against building a guitar. That would be the technological approach. Not to mention the fact that I would have no idea if my self-made guitar was playable or more useful as a decorative centerpiece for dried flowers. I should just buy a guitar, ignore the technology, and concentrate on the music and learning to play. Off to the local music store I went.

Entering the music store was a little like entering a gathering on another planet. There was a person with streaks of electric blue hair, strumming on a strangely shaped electric guitar, which was causing sounds that I would characterize as shrieks and rumbles to emit from monstrous speakers. I was between 20 and 40 years older than everyone in the store, although it was difficult to tell with some of the aliens. A friendly, more normal looking person asked if he could help me. "Yes," I said with more confidence than I felt. "I'm looking for a guitar."

"Acoustic or electric," he asked.

Feeling more confident because of my luthier study, I said "acoustic."

"Solid top or laminate?" he asked next.

My confidence fading, I looked uncertain.

"How much do you want to spend?" he asked helpfully.

I named a figure. He led me to a wall full of shiny guitars that he said were for beginners. A little annoyed at being pegged as a beginner so quickly, I picked up the nearest guitar and tried to give the appearance of evaluating the instrument critically. The salesperson launched into a rather technical discussion of how guitars make sounds. Some of the discussion sounded familiar from my luthier study.

In summary, an acoustic guitar amplifies the sounds of the vibrating strings by a coupling to the top of the guitar. (An electric guitar skips this step and just uses a magnetic pickup at each steel string and electronic amplifiers.) The guitar top needs to be as thin and as light weight as possible so that it vibrates easily. Lightweight and thin must be balanced by the strength needed to support one end of the tightly stretched strings. Of course, the strings are not really strings but copper-bronze alloy wires.

The strongest and cheapest way to make a guitar top is to use plywood. The trouble with this is that the glue used to hold the layers together tends to make it heavier and less responsive to the vibrational coupling. The sound is weaker and less rich. The better guitars use thin, solid cedar or spruce tops with braces attached to the underside to strengthen them. Even better guitars use solid wood for the sides and back as well to make smaller improvements to the sound. The salesperson went on with a discussion of how the neck of beginners' guitars are glued to the body, while better guitars use dovetails. He described the use of bi-directional truss rods in the neck, and on and on.

The salesperson finally asked me why I decided to take up the guitar. I named traditional, inexpensive, portable, low technology and easy to learn. This is after I had plunked down my credit card for an intermediate-price guitar. "Well," the salesperson said, "Guitars are traditional and portable, but to make them portable, you will need a case. Let me talk to you about cases...."

Why Wheels Are Round and Other Mysteries of Simple Machines

The new set of tires I bought recently seemed to cause more vibration than the old set. The dealer tried several times to fix the problem and finally agreed that one of the new tires was only sort of round. This started me thinking about the wheel in general. Who decided that wheels should be round, and when was that?

We sort of know the answers to those questions. The wheel seems to have appeared in Mesopotamia, located where parts of Iraq and Syria are today, in about 3000 BC. The first application was not transportation but rather a potter's wheel. The good idea, or the wheels, may have traveled because wheels of various sorts appeared in China and ancient India around the same time.

So, why are wheels round?

A wheel is one of the so-called simple machines, of which there are several. In fact, there is a wonderful, but mostly useless, argument about how many simple machines there are. The traditional candidates are: the wheel, the inclined plane, the lever, the screw, the wedge and the pulley. Some people throw in a gear as number seven. There are arguments for three, five, six or seven. I vote for three, as you will see in a minute.

A wheel meets the main requirement of a machine in that it provides something called "mechanical advantage." In other words, you probably can't carry hundreds of pounds of stuff on your back or even drag that much stuff across the ground. With a simple cart on wheels, you can easily move hundreds of pounds. In a cart, the wheels allow you to exert a much smaller force to move a heavy load than if you just tried to drag the load. The wheel does this mainly by managing friction. The wheel converts the large sliding area of the load being dragged to the much smaller sliding area of a wheel turning on a shaft. This only works because the wheel is round and can roll.

The wheel can also provide a different kind of mechanical advantage when it is fixed to a shaft rather than turning on the shaft. In this case, a larger diameter wheel turning a smaller diameter shaft gives significant mechanical advantage. This arrangement was often used with large water wheels turning heavy grain grinders to make flour. This same type of mechanical advantage is used in pulleys and gears. This is why I don't consider pulleys and gears to be simple machines. They are made up of wheels.

The remaining simple machine candidates are the inclined plane, the screw, the wedge and the lever. The inclined plane is commonly used as a ramp. Pushing something up a ramp is easier than lifting it. A wedge is nothing more than an inclined plane turned on end. A screw, to me, is nothing more than a spiral inclined plane or a hybrid of a wheel and an inclined plane. So, that leaves the wheel, the lever and the inclined plane as the simple machines.

That brings me back to my tires. Of course, wheels have only been in North America since the Europeans came here. That's probably why we still have a little trouble making them round. Another couple of thousand years, and we'll have it.

The Steely Grip of the Ancient Blacksmiths

This past weekend I was collecting experimental data on the relationship of how good you feel to the number of hobbies you have. Specifically, I was trying to harden a small steel pin that was needed in a metal working project. The process was simple enough. I heated a piece of tool steel with a torch until it was cherry red and stopped being attracted to a magnet. Then I stood back and cautiously dropped it in a can of oil. Presto! Very hard steel. During this process, I felt a connection with my ancestors. (Maybe it was just the hair on the back of my hand that I singed.)

Working with iron is one of the oldest skills known to mankind. Only chipping stone and working with copper and bronze (copper-tin alloy) is older. The first iron was heated and hammered (forged) into the shape of spear points and other tools about 4000 BC in ancient Egypt and Samaria, using stones as hammers. This was well before the process of extracting iron ore from rocks (smelting) was developed, so the source of the iron was the occasional iron-nickel meteorite that these early folks found. The guy with the iron-pointed spear fashioned from a flaming meteor from the sky probably had rock star notoriety.

Beginning in 3000 BC, smelted iron began to appear in small objects. These objects are distinguished from meteoric iron by the lack of nickel that is always present in metallic meteors. The smelted iron objects were still mostly ceremonial because the metal was much more valuable than gold.

At widely different times in different parts of the world, the so-called Iron Age took hold. Copper and bronze were gradually replaced with iron. At first, the iron was mostly a low-carbon form. The spongy iron obtained by smelting iron ore was repeatedly hammered and folded to remove impurities (slag) left from the rock. The process also tended to remove most of the carbon. We would call this wrought iron today. The material was useful but would not maintain a sharp edge for weapons and tools. First in the Middle East and later in other parts of the world, blacksmiths discovered that iron heated for long periods with charcoal and quenched rapidly in water or oil was hardened. Although not understood at the time, this process produced an alloy of iron and carbon, which we would call steel, on the surface of the piece. Steel could be sharpened to a fine, persistent edge.

Over centuries, blacksmiths learned, by trial and error, how to control the hardness of the steel layer that could be formed on an iron piece. The sword makers, like the famous Japanese Samurai sword makers, became experts. Hardness is key to creating a keen edge, but very hard steel is so brittle that it will actually shatter like glass. At the other end of the hardness scale, steel is somewhat soft but quite tough. If steel is heated to an intermediate temperature lower than the cherry red critical temperature and cooled suddenly, the hardness can be adjusted to a range of intermediate levels. This very tricky process is called tempering. A Samurai sword has a hard edge, tempered to prevent chipping, and a softer body for toughness.

Today, metallurgists tell us that rapid quenching of carbon steel traps a particular, microcrystalline structure called martensite that gives hardness. Tempering dissolves some of that structure. Steel is now mostly worked in factories by machines. The few remaining blacksmiths mostly create artistic objects. Still, a piece of glowing metal seems to take me back a few thousand years.

Stitching Together Some Technology

The home sewing machine is a typical, but still interesting, technological story of the advancement of a useful device.

Sewing itself is an ancient art. More than 20,000 years ago, man used needles made of bone or horn and thread made of sinew to fashion garments. The development of a mechanical device to do sewing began in the late 1700's and early 1800's.

There were many sewing machine inventors, many patents and many lawsuits in the mid-1800's. The commercial winner in all of the legal churning around 1850 was a sometimes actor, sometimes inventor, and part-time bigamist named Isaac Singer.

Singer effectively integrated technology from several inventors to make a fairly simple, robust machine. Even though he lost some lawsuits from the original inventors and had to pay royalties, Singer's name became synonymous with sewing machines. As is usually the case with technology, Singer was successful because of the business practices he used rather than the technology itself. His success could be attributed to three practices he pioneered: 1) He partnered with multiple patent holders to form a patent pool so that they would not waste their time and money suing each other. 2) He implemented what we would call a time payment plan so that most anyone could purchase a machine without an outlay of cash. 3) He targeted a mass market, the home, rather than just tailors and other professionals.

Throughout the rest of the 1800's and into the 1900's, the Singer sewing machine became one of the most common household mechanical devices. Initially foot-powered with a treadle, these machines were converted to electricity by simply hanging a motor on the side.

Up to the 1960's most homes did some level of clothing making or repair centered on a sewing machine.

Most high schools had a home economics course that included sewing. My mother always called her sewing machine simply "the machine."

After the 1960's, things began to change. Mass produced clothing, including foreign manufactured clothing, began to mean that it was often cheaper to buy clothes than to make them or discard them rather than repair them. Women began more frequently to have careers outside the home. Sewing machines began to move to the attic. Mothers stopped teaching their daughters to sew. Classes in high schools shifted more toward academic and business skills rather than home skills. Fabric and sewing supplies stores and sewing machine manufacturers began to struggle. Home sewing was on its way out.

Now it appears that things are changing again. This time the change is not driven by economics so much as by artistic flare and self-expression. People are more interested in customizing their clothing and even making their own again. Time is still valuable, but now sewing is an artistic pursuit instead of a necessity.

Technology even has a role. Instead of dragging the old sewing machine out of the attic, people are buying new ones. Even a low-end sewing machine today will do a wide variety of intricate stitches. High-end machines are computer driven marvels and allow unlimited exotic stitches for decoration. Both low- and high-end machines allow beginners to produce impressive, artistic projects. Fabric stores report renewed interest in courses and materials. Singer and other manufacturers are seeing increased sales. The "machine" is back.

Since I have never successfully threaded a sewing machine on my own, I wonder if I could thread one of the new computer driven models. Maybe you just drop the thread in a bin, type "sew on top button, shirt," and stand back. I suppose it's more likely that it will talk to you in a computer voice: "It's very easy. You just wind the thread over this thingamajob, hook it under this dohicky, wrap it twice around this...."

Will the Last Standard Kilogram Stand?

When I studied the metric system in high school, the fundamental units – the meter for length and the kilogram for mass – were actual chunks of metal carefully stored in Europe. The meter was a platinum-iridium alloy bar with scratch marks exactly one meter apart stored in a particular configuration at 0 degrees centigrade. The kilogram was a platinum-iridium cylinder defined as having a mass of exactly one kilogram.

Almost immediately, these two standards became a problem. Because they are not very portable, they must be used to make secondary standards, which are used to make other standards, which ultimately produce metric rulers and scales that get used. The very act of handling these standards and secondary standards makes small changes in them, even under the most elaborate protective protocols. As length and mass measurements became more precise, these standards became more of a concern. With the advent of nanotechnology, materials in sizes around a billionth of a meter are common and need to be measured.

By 1960, the length standard had been replaced by a standard related to the wavelength of light from the heated element krypton-86. In 1983, this standard was replaced by the distance traveled by light in a vacuum in a particular time period. This is the standard in effect today. Although it doesn't sound simple, the meter can now be measured to any desired accuracy and used to make other standards in any reasonably well-equipped laboratory in the world. The speed of light in a vacuum is believed to be one of the fundamental constants of nature, and time can easily and precisely be measured. The meter standard meets all of the requirements for a good standard.

So, what was done about the platinum-iridium kilogram in the Paris vault? Unfortunately, nothing yet. A mass standard based on something other than a physical artifact has proven to be very difficult. Both the International Bureau of Weights and Measures in France and the National Institute of Standards and Technology in the U.S. are working on alternatives. So far, these organizations have not been able to produce a standard with all of the requirements.

In principle, a mass standard should be easy enough. Atoms of a particular element have accurately known masses. All you need to do is count out a specific number of atoms of an element like gold or platinum, and you have a standard that will be accurate to the mass of one atom. There are, in fact, laboratory procedures for handling single atoms. The problem lies in handling enough atoms to make a practical standard and still having an accurate count. Stay tuned. In the meantime, it's still the cylinder in the vault in Paris.

Because the US has still not really embraced the metric system at the ordinary person level, what about standards for the foot and pound? Shouldn't we be concerned about those standards?

Before continuing with the pound, I will digress a moment about the confusion surrounding mass and weight. Although technically quite different, the terms mass and weight are often used interchangeably and sometimes have the same units of measure. This comes about because of the way mass is measured here on Earth. When mass, the amount of matter in something, is placed in a gravitational field, the mass is acted upon by a force (the force of gravity), which we call weight. So, a mass of one kilogram weighs one kilogram on a metric scale in Earth's gravity. Someplace in space, a mass of one kilogram could be weightless or could weigh more than one kilogram. Technically, metric weight should be expressed in units of force, called Newtons, but no one ever does that.

The avoirdupois pound, invented by London merchants in 1303, is the pound measure that we still unofficially use, along with several other English speaking countries. The foot is still our favorite measure of length. Is there some platinum-iridium cylinder that is defined as one pound? Do we have King Henry's boot preserved in a vault somewhere for the standard foot?

Actually, the avoirdupois pound has been defined since 1959 as exactly 0.45359237 kilogram. No matter who defined the original foot, it is now defined as exactly 0.3048 meters. Now both of our favorite measures are tied to the kilogram and the meter.

I am glad I don't actually have to work with nanotechnology. If my ruler is off a little either way when I do woodworking, I can compensate with sandpaper or putty.

A Horse, a Horse, My Kingdom for a Horse

You may have believed, as I did, that the 19th and early 20th centuries were dominated, energy wise, by steam engines and locomotives, followed by the internal combustion engine and automobiles. A closer look indicates that energy from steam and transportation by railroads were important, but horses were the critical element to connect everything together. Horses brought people and freight to and from train stations. Without horses, the railroads would have been ineffective.

Horses crowded the docks when a steamship was in port. A portable steam engine of any sort was moved from one place to another by horses. People used horses to go to work, to buy goods in town or do heavy farm work. Police and fire departments were horse drawn. A good horse or two could be the most valuable property in a family.

The importance of the horse shows in our language. American English is still filled with horsey terms. Don't look a gift horse in the mouth. That's horse trading. Don't put the cart before the horse. That's a horse of a different color. There's no point in beating a dead horse. He's a dark horse. She's feeling her oats. Hold your horses! Get off your high horse. That's straight from the horse's mouth. You eat like a horse! You can lead a horse to water, but you can't make him drink. They live in a one-horse town. Don't change horses in midstream. Stop the horseplay! She closed the stable door after the horse had bolted. He's backing the wrong horse. Wild horses couldn't keep me away! Got a burr under your saddle? The list goes on.

The early automobiles were pricey, rather unreliable, and useless on muddy roads. Motorists often depended on horses when their machines broke down or got stuck. For the first years of the automobile, they were toys of enthusiasts. Car owners were also horse owners.

The horse is not a recent addition to mankind's sources of energy. Horses appear in cave art dated to 30,000 years ago. Most likely, these ancient horses supplied dietary needs rather than being harnessed, however.

Horses were probably domesticated as early as 4500 BC, although very early evidence for domestication is not easy to obtain. Clues, such as intact horse skeletons buried close to human remains, sometimes with scraps of materials that were probably tack, have been found. In some cases, horse remains were found in areas where there were no native horses. By 2000 BC, domesticated horses had spread over the entire Eurasian continent, and the evidence is clear. Often horses were buried with the chariots they pulled and in the tombs of their owners. Details of how the domesticated horse developed are cloudy. Only one undomesticated, ancient horse survives today: Przewalski's Horse in Mongolia. This horse is apparently not in the lineage of modern horses, however.

Over time, two basic types of modern horses developed. The stronger, heavier bodied work horse and the lighter, faster breeds like the Arabian and, more recently, the Morgan, the Saddlebred, the Thoroughbred and the Quarter Horse. Often members of some of these breeds are prized as race horses. The progress of man would have been severely hampered if some plague had permanently wiped out the horse.

Opinions vary on exactly what caused the transition of the horse to its recreational status today. Certainly a new generation of automobile owners and automobile companies pushed for better roads and better vehicles. The mass produced car came within the economic reach of most people. The horse gradually gained the reputation for being dangerous and dirty in a crowded city environment. At the same time, the internal combustion engine improved and began to replace steam engines. Factories increasingly switched to the electric motor to run equipment. Finally, even the railroads suffered from the competition of high speed motor freight and air transport.

Never a horse enthusiast, I began to wonder when gasoline was four dollars a gallon.

1910 to 2007: Just a Few Little Changes

During travel to my father's burial in a small town in Illinois, I had lots of time to think about the astounding changes in technology that have occurred since 1910, the year he was born. Growing up, he drove a horse, named Pansy, and wagon to town to buy supplies or sell a load of tomatoes to the canning factory.

By 1915, the richest man in town owned a car, and it was a sensation. He took people for rides for a nickel. My father apparently had a nickel, and it convinced him he would have one of these mechanical marvels someday.

Airplanes had open cockpits and were generally only seen in a small town if the pilot had engine trouble and had to land in someone's field for repairs. Planes and cars required frequent adjustments and repairs, and no license was required to drive or fly.

My father's family had kerosene (called coal oil then) lamps for light and wood for cooking and heat. A hand dug well and a hand pump supplied water. The pump needed priming to wet the leather pump seals. The outhouse was out back.

As a teenager, my father built a crystal set to receive some of the early radio stations. A store in town carried the materials to make the radio. A crystal set uses a galena crystal (naturally occurring lead ore) to rectify the radio frequency signal from the antenna and separate the audio signal. The user needed to hunt for a sensitive spot on the crystal with a tiny wire called a cat whisker. He delivered the local newspaper to earn money for the radio. A little later, his parents acquired a console radio for the living room.

In 1930, my father had three priorities. First, he had just finished courses in a business college in a neighboring city, and he needed a job. He was soon hired by an animal feed manufacturing company as a stenographer to take dictated letters using a now mostly vanished written language called shorthand. He typed the letters from his shorthand notes. All business correspondence was done by letters through the US mail, except for a few time sensitive matters that required a Western Union telegram.

Secondly, even though he had ridden the streetcar to classes and then to his new job, his girlfriend was back home – about an hour away by car. A car was what he needed. The home town car dealer made him a personal loan for a new 1931 Ford Model A.

Lastly, with a job and a car, it was time to get married. I, therefore, probably owe my existence to his success with a job and a Model A.

The depression years shaped many of my mother's and father's lifetime attitudes, as it did for a whole generation of people. The necessity to be both practical and thrifty followed both of them for the rest of their lives.

The business office of today bears little resemblance to the 1930's. Administrative assistants have replaced stenographers. Correspondence is largely email and faxes. Computer workstations have replaced typewriters. The same workstations have replaced office boys and pneumatic tubes for intra-office communication.

Cars are fuel injected with most systems computer controlled. There are no adjustments available to the driver. Planes can fly at 33,000 feet with air conditioned, pressurized cabins at 600 mph in almost any weather. Orbital space flight is almost not newsworthy, and men have been to the moon. We have robots on several planets.

The changes are too numerous to count. There are many lessons to be learned from my father's generation. He would never throw away something that could be fixed or used by someone else. He remained just a little skeptical about new technology, except for cars. That nickel he spent as a kid hooked him for life. When asked if he would like to have another 1931 Model A, he replied, "I like seeing them, but they were hard to steer, hard to start and rough riding. I like new cars."

Index